A BIOLOGY
FOR DEVELOPMENT

A BIOLOGY
FOR DEVELOPMENT

François Gros

Preface by Jean-Michel Roy

*Translation by Gill Ewing, Daniel Jones, Marcie Lambert
and Rachel Stella*

EDP
SCIENCES

17, avenue du Hoggar
Parc d'Activité de Courtabœuf, BP 112
91944 Les Ulis Cedex A, France

Printed in France

ISBN : 978-2-7598-0402-3

SUMMARY

A biology for development

PREFACE

The rhythm of scientific progress varies. Times of stagnation or even setback are followed by stages of rapid improvement when old obstacles seem to disappear as if by magic, suddenly laying bare to the researcher's curiosity a whole new area of feverish conquest. But these periods of keen exploration are marked by a certain confusion: attention is dispersed in different directions, the accumulation of results seems incoherent, discordant explanations of the same facts multiply, and predictions increasingly diverge. Thus, they are also periods when regular assessments must be made to take stock of the progress accomplished. The scientific community feels a particular need to understand how it was led to the present situation, to accurately register all genuinely new possibilities, to establish an inventory of the products of its multiple investigations, to assess their potential in terms of applications, and to measure the full extent of the remaining difficulties to be overcome in conquering the emerging world of knowledge. This task is by no means secondary to the process of scientific discovery. Quite the contrary, it is an essential one serving to clarify and guide this process, fostering its momentum by providing it with firmer foundations. Mapping a discipline in the process of transformation is still taking part in its transformation.

Biology, broadly understood as all the disciplines that take living organisms as their object of inquiry, is undoubtedly one of the main scientific fields to have entered such a phase of acceleration since the middle of the last century, as illustrated in particular by the revolution of molecular biology, which penetrated the deepest foundations of life by elucidating some of the most elementary operating principles and structures of the cell. That the unique impetus then given to biology has retained all of its strength today, is undeniably confirmed by the constant results flowing from laboratories, results sufficiently important to be echoed almost daily in the general press, provoking a mixture of admiration and dread in society confronted with the new prospects thus open to humanity. A single, but particularly striking, example demonstrates this vitality. To improve understanding of embryonic stem cells, researchers in Great Britain recently received authorisation to produce embryos of indeterminate status, since these embryos are obtained by the insertion of an adult somatic cell nucleus into an animal oocyte, thus carrying a human genetic inheritance without however qualifying as human ones. By making it possible to overturn one of the most fundamental divisions of the ordering of organisms elaborated by evolution, the scientific exploration of life cannot offer better proof that it has well and truly crossed the frontier of a new territory, the limits of which remain impossible to determine today.

If biology is consequently also one of these scientific fields where the need to take stock of its progress is most pressing, clearly no-one is in a better position to perform this arduous task than biologists themselves, at least those who can fully appreciate the extent of the transformations affecting their discipline as well as master its main theoretical foundations.

Such is precisely the task that a renowned geneticist addresses in this book, even as he emphasizes the benefits that this blossoming of biological knowledge might bring for the resolution of the challenges raised by the development of contemporary society, and also the perspective offered by genetics which is his field of expertise. Few people are indeed better qualified for this undertaking. Not only was François Gros a key player in the molecular biology revolution, through his work on messenger RNAs in particular. He is also one of its most accomplished analysts and historians. From *Secrets du gène* (1969) to his *Mémoires scientifiques* (2003), via *La Civilisation du gène* (1989) and *Regards sur la biologie contemporaine* (1983), he has repeatedly strived to bring to light the theoretical significance as well as the societal implications of this dramatic transformation in biological knowledge, in which he was both an actor and a privileged witness at the side of Jacques Monod and François Jacob. Nearly thirty years after the report he wrote with Jacob and Pierre Royer for the French President about "the consequences that the discoveries of modern biology might have on the organization and functioning of society" (*Sciences de la vie et société*, 1989), he offers in this new book his latest analyses on how the most recent advances of biological investigation may, and within what limits, lead to significant progress in the domains of health protection, food production and energy supply, three essential areas of social development. Furthermore, by first recalling the main episodes of the scientific adventure which led to these advances, he sheds light on the relation between the trials and errors of basic research and the applications that they ultimately make possible, the way in which basic research fosters potentialities that applications only gradually delineate, at times so unpredictably that the original discoveries from which they derive appear even more fabulous.

François Gros' reflections are not only of interest with regard to the actual state of biological knowledge today, but naturally also to the issue of development and the means of resolving it. From this point of view, their significance is even greater when we read them in the light of the new relationships being built between science and development.

Indeed, these relationships have changed. Society no longer asks science simply to help it develop, but also to help it develop differently. Society no longer expects from science only more development, but also another type of development, thus granting to scientific knowledge a more essential role than ever in the quest for a better future, and increasing in similar proportions the responsibility of scientists. To fully appreciate the extent of this new responsability requires one to briefly examine the close relationship that has always united development to science, and the unprecedented crisis in which this relationship unquestionably foundered over the last quarter of the 20th century.

But it should first be asked what people exactly mean when talking of the development of societies. Although it is now recognised as the subject of a specific discipline known as development theory, with its own research centres and training courses, as well as a myriad of national and international institutions with a more practical aim, this notion is still seldom defined. Even if it clearly refers in all its uses to a process of transformation, it is not unequivocal, and when applied to societies, its meaning is rather specific. It would be particularly inappropriate to take it as synonymous for growth in such a context. For even if an increasing population is admittedly a population which in a certain sense of the word is developing, this growth is often a factor of under-development. Similarly, a transport system perfectly identical to another one in terms of geographical points connected, number of vehicles in circulation and amount of people transported, but much more advantageous in terms of energy consumption and toxic emissions, would clearly be considered by development theory as more developed than the other one. The concept on which this theory is based is therefore also, and even primarily, qualitative. It seems reasonable to suggest that, at its most general level, this concept refers to the process by which a society improves its global capacity to satisfy the needs and desires of its population. And, consequently, that a difference in degree of development corresponds to a difference in state of such a capacity.

This definition is particularly recommended for its aptitude, through the idea of global capacity, to explain the fact that a society considered as having reached a certain degree of development may, in reality, display major differences in its capacity to meet each one of the different needs of its population, for example between those of health and those of transport. It also gives the notion of development a welcome relativistic flavour without lapsing into relativism. For, if the degree of superiority of a capacity to satisfy certain needs with respect to another can largely be determined by objective criteria, what is considered as developed at one stage is nevertheless destined to look underdeveloped at another, unless any improvement is impossible. Moreover, it leaves entirely open the possibility that some ancestral technique, for example in agriculture, should be estimated better than a modern technique when ecological impact criteria are taken into account, and therefore that the ancestral one represents a higher degree of development. Finally, this way of looking at development seems to fit its human specificity. Indeed, it is patent that animal societies, in the sense of development theory, do not develop. Only human societies do, because animal societies are deprived of the ability to truly improve their capacity to satisfy their needs and desires. Ever since they came into existence, gazelles have found no better way of avoiding ending up in the lion's digestive system than trying to run faster than him. At best, an animal moves from one territory to another to make the most of the capacity to meet its own needs with which it is naturally endowed, but it cannot improve this capacity itself, and transform its condition as a result. The human societies that we call primitive can be considered as societies engaged in a particularly slow process of such improvement, or in a process of improvement different from the one that has become predominant in the world through Western Europe.

It seems undeniable that certain types of political, legal and economic organisation within a given society, or in its external relations, can contribute directly to development so understood. However, it is no less true that these organisational factors have above all a crucial role to play with regards to guaranteeing fair access to the improvement process that defines development. This improvement process itself is rather a matter of advancement of knowledge, not of the kind of metaphysical speculation or religious revelation, but of the kind we call scientific and technical, the latter generally being based on the former. It is in fact this decisive contribution to improving the capacity of societies to satisfy their needs which has ensured the success of the scientific enterprise since the 17th century Scientific Revolution; and it does not seem an exaggeration to say that it has been gradually converted entirely to its service.

Although it always had critics, the trust thus placed in the power of science to help societies make progress towards satisfying their needs and desires was shaken in a particularly deep way in the last three decades of the second millennium. This disruption originates in the realisation that the progress made has, in fact, been accompanied by serious negative effects, such awareness being itself largely the consequence of the increasing amplitude of these effects that came along with the acceleration of progress. In other words, development reached crisis point at the same time as it reached its peak, precisely because its drawbacks themselves reached an extreme state. This is why this crisis, in a way the price of success, has been concomitant with the unprecedented industrial transformation of the period following the Second World War, marked both by the rebuilding of Europe and the progressive industrialisation of Third World countries, thanks to the decolonisation process.

It finds a very clear expression at the level of international institutions, and of the United Nations Organisation in particular, even if the attitude of these institutions owes a great deal to the critical work of a number of pioneering individuals and associations. It is for instance under the influence of the warnings launched in 1970 by the members of the Club of Rome in their famous report *Halt on growth*, that the UN formulated its first major concerns and recommendations during a conference on the environment organised in 1972 in Stockholm, where it was solemnly declared that "a point has been reached in history when we must shape our actions throughout the world with a more prudent care for their environmental consequences", and that "through ignorance or indifference we can do massive and irreversible harm to the earthly environment on which our life and well being depend." The first consequence of this declaration was the creation of the influential *United Nations Environment Program* (UNEP), then the establishment of the *World Commission on Environment and Development* which conducted a survey from 1983 to 1987 on the state of development on the planet. Headed by Gro Brundlandt, a former Prime Minister of Norway, this commission provided in particular an explicit definition of an alternative development route which the UN invited all countries in the world to follow, and paved the way for the most symbolic of these large international meetings which gradually called into question the opposite path followed until then: the 1992 Earth Summit of Rio de Janeiro, which, among a number of important resolutions, approved the famous programme of measures known as Agenda 21. The *United Nations Commission on Sustainable Development* (UNCSD) was created at the same time to monitor the implementation of this Agenda; and in 2002 the latest UN general summit on development was organised, the *World Summit on Sustainable Development* of Johannesburg, to assess the progress made and give new impetus to the commitment of nations.

It is important to highlight the singularity of the criticism of development, and of the scientific progress underpinning it, which drives this whole movement. Indeed, a full tradition of political, sociological and philosophical thinking had already thrown suspicion on them a long time ago. But the motivation of this tradition was, above all, to denounce inequality in the distribution of benefits which development and scientific progress had provided and could provide, as well as the pillaging of the resources and labour of a portion of humanity on which they were seen to rely. Or also, at a later stage,

to denounce the cultural impoverishment that they brought through the establishment of a consumer society centred on the search for purely material well-
being. The criticism which took shape in the early 1970s in the international
movement described in the previous paragraph is much more radical, since
it questions the very ability of development and scientific progress to ensure
material well-being, in the name of the new threats they engender, and which
are considered as powerful as the ills they are supposed to overcome. Hence
the decisive importance played in the emergence of this criticism by the question of environmental damage, that provided the earliest and most explicit
manifestations of these ills. An importance well reflected in the chronology
of the UN response to the crisis. The Stockholm conference which marked its
starting point was indeed a conference on the environment, and resulted in
a declaration about the environment. It was only in a second phase that the
Brundtland commission emphasized that environmental issues could not be
separated from those of equal distribution of the fruits of progress, basing
their argument on the premise that progress, in order to avoid being self-
destructive, must not destroy the environmental capital from which it draws
its source, and should also be able to distribute dividends to the entire human
population. The role played by climate change is equally meaningful in this
respect. Through their global and no longer local dimension, the increasing
manifestations of climate change over the last few years clearly marks the peak
of this entire crisis, as well as at the moment when it grips the imagination on
a global scale and ceases to be the sole affair of committed ecological groups,
thus demonstrating unequivocally how much its source lies in the damaging
effects of development on the environment. What good is it to have factories
supplying us with any number of consumer goods or with vehicles which can
take us anywhere we wish, if the price to pay for them is a climatic disturbance
which dries or floods the land on which we live and which feeds us, making
us vulnerable to apocalyptic physical threats and famines man has fought to
overcome over the centuries?

By questioning the ability of development and scientific progress to
improve our material well-being, this criticism is also more radical in that it
questions their reality itself. Not by denying their undeniable successes, but
by invoking the ills which these successes have engendered, counterbalancing
them to the extent of destroying them at times. The basic idea behind such
criticism is therefore that the process of development undertaken by humanity so far, having "reached a certain point in its history", has started to cancel
itself out. That its apparently eternal victories are actually only temporary,
either because they have simply eliminated one evil only to generate another,
or because what they seemed to have vanquished is rising again from its ashes,
like these biblical floods or food shortages which once again are thought to
threaten the 21st century.

And this is also why the fundamental claim to which it gave rise is that of sustainable development. What is indeed a sustainable development, if not a development which is perennial because it does not cancel itself out in time? It would be a mistake, however, to see this notion as a radical innovation. Apart from a few excesses, development has always been essentially intended to be sustainable and to produce permanent benefits. The problem is rather that it has proved not to be so, and that we can no longer delay replacing it with one that really is. The form of development followed until now can only be qualified as non-sustainable in the sense that it was under the illusion of being sustainable, and not in the sense that it was deprived of the intention of being so. Undeniably it can be criticised for falling under this illusion partly at least through heedlessness, and therefore owing its failure to thoughtless attitudes. This is why also sustainable development can only be conceived as a form development truly guided by a new intention if it is understood as one animated by a much deeper concern for its sustainability. The demand for sustainable development is not a demand for a development with no risk of cancelling itself out, but for a development which takes every possible precaution to minimise this risk. It is in this only that it constitutes a new mode of development, whose novelty is nothing else than a much more vigilant awareness of its responsibilities.

But how can this be achieved? The temptation is always great, when faced with a sizeable failure, to challenge that which led to it. This is a perfectly rational attitude, unless it lapses into excess. Can we relinquish scientific knowledge and break its secular link with development? Our society is clearly incapable of it, apart from the question of whether a society can in principle develop without science. This is no longer a real possibility for ours, even if it were one in principle. What the Stockholm declaration probably does not emphasise enough is that the point we have reached in history is also, in this respect, a point of no return, and that we are condemned to maintain the alliance of development and science. First of all, because science alone can adequately diagnose the damage already done, as shown by the numerous expertise agencies nowadays needed to provide guidance about the purity of our water, the harmlessness of our food and the innocuousness of our air, and of which the *Intergovernmental Panel on Climate Change* is perceived as the heroic figure. Secondly, because science is essential for repairing much of this damage, even if it cannot do it all. Finally, because science alone can supply most of the instruments necessary for the survival of the humanity born out of our past, unless we accept to plunge it into an unprecedented regression, of precisely the sort which the club of Rome was predicting under the name of collapse. The development crisis condemns science much less than it condemns us to science. Science must continue its construction, renew its efforts and go beyond its present limits, in order to repair past mistakes it made possible as well as to open safer horizons. Something which cannot be achieved without crossing new frontiers. Society expects for instance science to reduce the excessive concentrations of atmospheric carbon dioxide resulting from the exploitation of fossil fuels, to replace these with a cleaner and yet as efficient energy, and finally to move forward in the direction of energetic abundance. A threefold expectation which cannot be fulfilled without penetrating natural mechanisms of which we know very little at the present time. The challenge of sustainable development is essentially a challenge of scientific and technical innovation, because it requires from us that we can do things currently beyond our capacities, and to which only research can give us access.

    This primary role of scientific knowledge in the conquest of sustainable development is widely acknowledged and has been reasserted with growing force, for over almost forty years now, at each one of the crucial stages of the mobilization in favour of sustainability. As early as 1972, principle 18 of the Stockholm declaration states that "science and technology... must be applied to the identification, avoidance and control of environmental risks and the solution of environmental problems and for the common good of mankind." And the Rio earth summit declares in chapter 35 of Agenda 21 that "scientists have a growing understanding of such issues as climate change, increases in resource consumption, population trends and environmental degradation... [which] should be used to shape long-term strategies for sustainable development", and officially recognises the scientific and technical community as one of the nine social groups essential to the reorientation of our mode of development. As a result, this community was closely associated to the preparation and realization of the Johannesburg summit through the *International Council for Science* (ICSU) and the *World Federation of Engineering Organizations* (WFEO). Finally, the Nobel Prize awarded to the IPCC in 2007 obeys to the same line of thought, and at the same time brings it to culmination. But it is perhaps in the presidential speech given by biologist Jane Lubchenco at the annual congress of the *American Association for the Advancement of Science* of 1999 that this recognition of the primary role of scientific knowledge found its most vibrating expression. Acknowledging explicitly that "fundamental research is more relevant and needed than ever before", she called for the entire scientific and technological community, on the threshold of the third millennium, to sign a "new contract" with society, in which it agreed to "harness the full power of science of the scientific enterprise... in helping society move towards a more sustainable biosphere".

    This way of using science to serve the pursuit of a mode of development concerned about its sustainability is not without implications on the way scientific investigation should be conducted, and of which ICSU, in line with its participation at the Johannesburg Summit, has offered interesting analyses, particularly in its report on *Harnessing Science, Technology and Innovation for Sustainable Development.* This new approach to science calls in particular for the establishment of new forms of interaction between the scientific community and other components of society at different stages of the scientific process, in order to coordinate the choice of research objectives with the actual needs of populations, to allow the results to be more easily available to those who can exploit their industrial potential and those in charge of legislation and government, to foster better understanding of their applications by public opinion, and to discuss the ethical and legal issues as constructively as possible. The transformation of science required by sustainable development is not only a matter of discoveries, it is also a matter of ways of investigating.

And it is in this twofold perspective that *A biology for Development* should be read. Although François Gros' main purpose is to clarify what the major results of contemporary biology, generated by its molecular revolution, are starting to bring to the quest for of a form of development at long last in control of its consequences, his commitment to greater interaction between the world of science and the rest of society in the process of research is unambiguous. Not only does he support the work of the ICSU, to which he refers, but throughout his career, especially through his responsibilities as head of the Pasteur Institute and the French Academy of Sciences, he has multiplied such initiatives. His decisive involvement in Biovision, the World Life Sciences Forum, is perhaps one of the clearest illustrations of this commitment. Organised by the Scientific Foundation of Lyon, the Forum is precisely a decade long effort to contribute to a more efficient articulation of the life sciences with the main developmental challenges of contemporary societies, by inviting researchers, industry representatives, political decision-makers and opinion leaders to debate the most recent orientations of biological research in the light of pressing developmental issues. François Gros' book undoubtedly draws some of its inspiration from his involvement in this unique and daring enterprise. All those who designed, founded and animated it will surely find it to be a powerful stimulus for perseverance. And they thank him for this.

Jean-Michel Roy
University of Lyon
École Normale Supérieure Lettres & Sciences Humaines

GENERAL INTRODUCTION

One of the dominant traits, and doubtless the most characteristic of our entry into the 21st century, is the awareness of the evolution of the global environment, which the concept of sustainable development conveys rather well. Awareness of the planet's natural wealth, a kind of common heritage but which is going to run out (water being the resource par excellence), increasing preoccupation with the effects of global warming and the spread of emerging diseases, not to mention the rural exodus, urban overpopulation and particularly economic lagging behind of the most deprived countries which contrast with the new consumer appetites displayed by others.

Although this somewhat Promethean vision often responds to the self-protective reflex of large affluent countries, it also demonstrates, for the first time in history, a movement of genuine international solidarity in meeting all of these challenges by working together. That science has a role to play in managing the solutions of these enormous, very long-term problems, reflects an expression and even a dream which recalls the ideas of the 18th century. What is new is that we have moved on from discussions, prayers and good feelings to organised, concrete and concerted thoughts on how to act and according to what criteria. The millennium goals defined in Johannesburg bear witness to that.

This book intends to assess the current or foreseeable contributions of the life sciences as regards the general issue of development. Why life sciences? Because in their assets, methodology, and recent technical advances, they contain the answers to many questions of this century; sometimes long term, sometimes "within arm's reach". Through molecular and cellular biology, recombinant DNA techniques, genomics and bioinformatics, life sciences have actually gone beyond the inventory stage of listing and describing natural phenomena, however essential that may have been, and still remains, to become genuine "intervention science". For the best but also for the worst some might say! But human wisdom, a permanent and constructive ethic and a deliberate policy must succeed in giving precedence to what is "best" ...

In the first part, priority is given to the historical dimension which has accompanied the long journey of biological thought, from the particular vision of the ancients to the premises of molecular biology. A long journey in fact, during which biology gradually became an experimental science after having been the expression of metaphysics and dogmas, and finally found its unity in the study of cell life before defining a principle of universality of the living world in the reactivity of macromolecules... The major breakthroughs which molecular biology has achieved for us will therefore be reviewed to the present day when we are trying to move beyond the characterisation of genes and their products, to reach a genuine integrative biology or "systems biology". This deals with the enormous complexity of the "networks" of gene regulation and the physico-chemical interactions of proteins in the hope of explaining the major functions of life. With the extra benefit of bioinformatics, the nanosciences (DNA chips), high-resolution physical techniques and even applied mathematics and simulations, this post-genomic approach attempts to elucidate processes as complex as: evolution, reproduction, differentiation, cell ageing and death, not forgetting some of the chief functions of the nervous system.

In the second part, applications from contemporary biology and biotechnologies are illustrated, with the analysis, prediction and often even the solution to some major development problems such as: knowledge and protection of biodiversity (metagenomics), genetic diseases, cancers and the premises of gene therapy, stem cells and the hopes placed in regenerative therapies, the fight against infectious diseases (particularly the anthropozoonoses) and, in another aspect, the recent contributions of biology to sustainable agriculture (the fight against hunger, climate prediction, transgenic plants, socioeconomic and ethical aspects).

Biology is not the panacea which will solve these major issues which man and sometimes the entire planet face. The great importance of national as well as individual economic status, political endeavour and even the role of human and social sciences, are determining factors, the importance of which does not escape the author of this book.

But biology and its technical consequences, supported, it is true, by many other scientific disciplines, give us a better understanding of the nature and severity of the great challenges of development while often providing answers and even long term solutions. It would be a pity not to take them!

THE FABULOUS DESTINY
OF BIOLOGY

I.1. HISTORICAL OVERVIEW

I.1.1. THE VISION OF THE ANCIENTS

The history of biology is ensconced in a past so distant that accurately locating its beginnings is very difficult. It is hard to imagine that, since the dawn of consciousness, man has not tried to question the very essence of his being, such as the relationships between the physical universe and life in general. Several factors, even various "driving forces" of inspiration, must have been the cause of it; to start with, curiosity, this precursor of knowledge (to which allusion is made in the Bible!), the fear of great cataclysms and the search for their hidden meaning which should lead to supernatural dimensions; then religion or, more practically, the desire to classify living beings according to their proliferation. Later still the idea of domesticating plants and animals and therefore becoming more familiar with their distinctive features and properties would assert itself.

Through its independence relative to the inanimate world, the nature of life (as distinct from that of the physical world) aroused the curiosity of the ancients early on. For influential thinkers like Pythagoras (570-480 before J.-C.), Plato (428-348) or Empedocles (490-438) beings and things are composed of a mixture of four major physical elements: air, earth, water and fire, in proportions depending on their specificity. For Democrates, inanimate objects and living beings result from the assembly of atoms, but life would prove difficult to define for a long time; if it doesn't move it is not alive! This led to the idea that plants, for example, are only imperfect animals incapable of moving from one place to another, inverted animals, which have become immobile, their heads buried in the soil and whose roots were formerly hair! However, procreation, the ability to reproduce and grow as well as morphology, physical imbalances, and diseases of all types would gradually led to medical and surgical knowledge, resulting in the first milestones of anatomy and physiology and paving the way for greater knowledge of what we now call biodiversity.

In this respect, Aristotle (384-322) can be considered to be the first "naturalist" in history. We are indebted to him for having listed and classified over 500 species of animals, insects, fish, reptiles and mammals; he was also interested in their growth, their methods of reproduction and locomotion. The title of "father of botany" should be given to Theophrastus (372-287). He described hundreds of plants, proposed ingenious classifications, observed the germination of seeds and specified the therapeutic properties of a number of plant extracts.

The first dissections of the human body go back to the pre-Socratic period. The Greek doctor Alcmaeon (6th century BC) acknowledged the brain as the centre of thought which up until then was assumed to reside in the heart! He had the idea that health results from perfect harmony between all the "substances" of which the body is composed and that illness is therefore caused by imbalance.

While listing the theory of the four elements, Hippocrates also intro-duced the doctrine of the "humours" (more biological!): blood, yellow bile, black bile and phlegm! Their "health-illness" balance depends on their harmony, good balance as it were, which did not prevent Hippocrates from surmising that fire, or "inflamed air" is the first principle of life or "pneuma" which gives life to the heart and blood vessels. Then, from the observations of Herophiles (330-260) and Erasistrates until the systematic works of Galien, in the 2nd century AD, anatomy continued to progress, leading to the develo-pment of surgery and opening the road to on the physiological knowledge of the human body.

It would therefore be tempting to say that, little by little, "life" was in the process of finding its place in the universe with the first descriptions of the animal and plant kingdoms, thanks to increasingly accurate knowledge of animal and human anatomy. It could therefore be thought that the "continuist" vision of the first Greek philosophers (School of Millet), where the natural elements which make up the inanimate world are also the constituent elements of life, would give rise to the forms, content and even the mechanisms of what we now call the biological world. But this would be to jump the gun.

In fact, the history of antiquity and its extension until the beginnings of the Renaissance demonstrate that the mind is conststantly in search of "a permanent unchanging and <u>abstract</u> order from which we can deduce the changing world of observation" (A. Cameron). For Plato (428-348), then, perception was only an illusion. According to him, ideas were above objects, and it was essential to give supremacy to reason and to search for the reality behind appearances. Therefore abstraction, the invocation of natural and supernatural forces were often involved in the search for the causes of life. Like Galien's famous "pneuma", for example, or his "principle of heat". According to him, then, it was the "pneuma" which causes muscles to contract under the effect of nerve stimulation. To Descartes, these were "animal spirits". It was only much later (in the 17th century) that physiologists discovered the existence of reflex mechanisms. It was not therefore medieval ideas, where Greek thought and culture were often taken over by the Arabs but theology continued to occupy a prominent place, which would trigger the "move" from speculative theory and the abstract world of ideas to <u>experimental</u> verification. However the particular status of alchemy should be mentioned here[1].

I.1.2. THE NATURALIST EPIC AND EXPERIMENTAL PHYSIOLOGY

It was, however, the British doctor, William Harvey (1578-1657), during the period marking the transition between the 16th and 17th centuries, who deserves considerable merit for enabling life sciences to be acknowledged from then on as an experimental science, thanks to the discovery of the blood circulatory system. Through his works, particularly those focussing on the ligature of blood vessels, Harvey was not only breaking with the obscurantist thought which prevailed on the subject of the major functions of the human body, but he was a precursor of experimental physiology which developed later on, in the 19th century, and helped sketch the specific outlines of a genuine "life science".

From the 17th century and during the 18th, the life sciences experienced a major turning point. On the one hand, work on human anatomy and comparative anatomy (Fabricius of Acquapendente, Wharton, Sylvius, Pecquet and de Graaf) progressed rapidly. On the other, genuine "taxonomy" started to develop, illustrated by the classification of the plant kingdom, (with the

---

1. We know the importance of "metallurgical experimentation" (especially the use of sulphur and mercury) by the first Greek, Arab and Chinese alchemists. Admittedly, mysticism was not absent, far from it, from the use of liqueurs, sublimates and elixirs of all types (Geber 721-815). Nevertheless, we can see, in these practices, the beginning of experimental method. Alchemy actually came close to chemistry in its Western version (13th and 14th centuries) and particularly began to promote prophylaxis and medicine, with Bruno Valentin and Paracelse in the 15th century, who introduced the use of antimony salts and mercury for the treatment of infectious diseases (syphilis).

Cesalpino and Baudin brothers), and particularly the colossal work of Carl von Linnaeus (1707-1778), the originator of binomial classification, in the animal kingdom (Buffon, Daubenton 1716-1800).

The 18th century was the golden age for the first naturalists. Their attitude towards living things borrowed from the scientific approach during the first half of the century at least, by trying to extricate itself from the metaphysical dimension. Instead, it attempted to place the emphasis on observation techniques. Among the scientific criteria to which its most illustrious representatives (such as Bonnet, Necdham, Spallanzani, Buffon and Réaumur) claimed to adhere, the consistency of experimental data even before their reproducibility, figures prominently. The prevailing idea from then on was that without a long series of coherent observations possibly coordinated around a particular theme, they would not be considered as having done the work of a true scientist.

This naturalistic phase (which in reality began at the end of the 17th century, encompassing the first two thirds of the 18th), was, in actual fact, an undertaking to draw up a vast index of living beings. Thousands of animal and plant species were described and classified at that time. The number of species characterised in this way increased considerably. For all that, the naturalists' work was not limited to a simple inventory (vast as it was). It was the work of Charles Bonnet, Spallanzani, Réaumur and Buffon, that developed true physiology, i.e., the attempt to understand the major functions of life. Admittedly, and as we have seen, the English physician Harvey set the tone early on with his discovery of blood circulation, but we owe a multitude of information and at times highly detailed description of processes as numerous and varied as animal "procreation" (Ch. Bonnet), sexual reproduction (Spallanzani), limb regeneration in the aquatic salamander and polyps (A. Trembley, Bonnet), the effects of temperature on egg hatching (Réaumur), the life of bees and more generally, insect behaviour (Buffon), to this work of naturalists during the period preceding the French Revolution.

It was also the era of botany par excellence, herbaria and herbalism, so beloved of Jean-Jacques Rousseau! Here too, the vast plant kingdom began to be physiologically interpreted (the effect of light on plant growth, absorption of water by leaves, etc.). Faith in science had rarely been so strong! It was, however, towards the end of the 18th century, during the period following the French Revolution, (which saw the creation of the Natural History Museum (1793) replacing the Royal Garden of Medicinal Plants), that naturalists would show what they were capable of undertaking the first large all-encompassing approaches to life sciences. This led to some of the most fundamental hypotheses and results in biology, in its very essence, namely the corpus of ideas on <u>evolution</u>.

Buffon (1707-1788), keeper of the King's Garden, was one of the first naturalists to take an interest in the origin of species, by imagining that they were derived from "organic molecules". Mariotte, then Maupertuis (1698-1759) in particular, went even further, since the latter, in his "essay on the formation of organised beings", manifested the attitude of an evolutionist before the term was ever invented, with a corpuscular hypothesis of heredity "in which chance transformations fathered the diversity of living beings" (D. Buican in "Genetics and Evolution", Que Sais-Je? 2nd edition, 1993). But these attempts at an atomistic explanation of the origin of species, prefiguring the existence of genes and mutations, were ignored for a long time, because they were too advanced for their time. Mendel's Laws were only developed in 1863, going unnoticed and ignored for decades, to be rediscovered at the very beginning of the 20th century (Hugo de Vries, Correns, Tshermak).

It is also important to emphasize that the first major global theory of evolution, i.e. "transformism" and the inheritance of acquired characteristics, appeared in 1809, in J. B. Lamarck's "zoological philosophy", devised well before the birth of genetics. The famous holder of the Chair of "Invertebrates" at the Natural History Museum, although a supporter of spontaneous generation, through his experiments on infusorius, had the idea that much more complex beings are progressively formed from primitive organisms. He explained these gradual transformations towards the acquisition of characteristics leading to increasingly complex organisation through two mechanisms.

According to him one mechanism involves a kind of "inherent tendency" in living beings, even in "living matter", <u>towards improvement</u>! The other is due to the pressure of external circumstances: adaptation to the environment gradually modifies physiology and anatomy, etc. These modifications are transmitted through heredity, which explains the diversity of species.

Lamarck's tranformism would, however, come up against the "creationism" of Georges Cuvier (1769-1832). This great zoologist, the founder of palaeontology, established a genuine zoological classification. So, paradoxically, although his observations on extinct species could have supported Lamarck's transformism, Cuvier contested the latter's ideas, although he also championed those of his friend and colleague, Geoffroy Saint Hilaire. Creationism is a hypothesis according to which, species, in their present diversity, all appeared <u>as they are today,</u> through the will of a creator. In creationism, heredity is invariable. Biological diversity may also result from sudden changes in the environment and lead to the extinction of species whose existence had been desired by the creator (catastrophism), which explained the discoveries of fossil organisms of which Cuvier himself was a major specialist. Transformism gave rise to many variants and inheritance of acquired characteristics was prevalent for a long time (including in its common meaning!) until the publication of "The origin of species" (1859) and the dawn of Darwinian theories, that gave evolutionary theory the new insight we know about, and to which the modern scientific community generally adheres, namely, natural selection which is brought to bear on inherited, possibly random, variations. We know how, by postulating that the selective effects of the environment on living organisms during their reproductive stage and geographical isolation lead to the propagation of those which have acquired selective advantages, Darwin explained the origin of the diversity of species (without rejecting the inheritance of acquired characteristic).

We owe a debt to August Weissmann for having definitively refuted this Lamarckian mode of hereditary transmission through his experiments on mice and particularly for having formulated his theory of the continuity of "germinative plasma", due to the existence of "nuclear filaments" (future chromosomes) in reproductive cells, for the first time in 1883. Weissman suggested that Darwinian selection operates by acting on reproductive cells, and not, as Lamarck thought, on somatic cells.

Weissman's ideas would therefore give rise to the chromosomal theory of heredity which the great American biologist T. H. Morgan developed in 1926, in his "Theory of genes". Neo-Darwinism found its definitive expression once mutations were discovered (Hugo de Vries, Müller) and after establishing the "junction" between natural selection and the modifications generated during mutations at the genetic level (the synthetic theory of evolution)[2].

---

2. By anticipating with respect to the historical overview of the problem as developed here (i.e. those covering the period from the first evolutionists to the upholders of the theory of natural selection and the mutationists of the early 20th century), it can be said that modern evolutionary theories have been inspired by cytogenetics, molecular biology and genomics. The emphasis has been placed on processes such as chromosomal reorganisation (recombination, fusion, etc.) with maintenance of sexual fertility and on natural selection operating on a series of micro- or macro-mutations affecting development genes etc. Recent studies, focussing on the comparison of genomic sequences, have also elucidated the probable mechanisms of transition from one subspecies to another (yeasts); Teleost fish). We are also endeavouring to understand how the conjunction of genetic modifications and the selective pressure exerted by the environment (geographical isolation, transition to bipedalism , etc.), but also learning, etc., could have promoted the emergence of the "Homo" genus among the large anthropoid apes, , endowed with a brain with superior cognitive abilities and the faculty for language.

During the 19th century, the life sciences continued to develop with the description of major physiological functions, through breakthroughs in the field of embryology and the spectacular development of biochemistry.

Finally, following the work of physiologists such as François Magendie (1783-1855), and particularly Claude Bernard (1813-1878), vitalistic theories ceased to prevail. So, according to Magendie: "physiology, intimately linked to the physical sciences, could no longer progress without their help; it gained the precision of their method and language and the certainty of their results..."[3]. Claude Bernard, the author of "An Introduction to the Study of Experimental Medicine", developed the basic methodologies of modern research in the life sciences by clarifying the <u>principle of objectivity in experimental procedure</u> and basing it on essentially deterministic foundations.

I.1.3. IN SEARCH OF A UNIFYING FORMALISM OF LIFE (ENZYMES, METABOLISMS, BIOENERGETICS)

By continuing the work of evolutionists and the deterministic direction taken by physiologists, chemistry, then biochemistry,would mark the start of scientific contributions aimed at emancipating life from the vital forces and the classifying epic of the naturalists. With the synthesis of oxalic acid (1824) and urea (1828), two products of metabolism, Wöhler created the foundations of biological chemistry. "Living matter" became matter, full stop! But biochemistry in particular provided the first explanations of life on the cellular scale with the enormous contribution of enzymology. Life results from a remarkable set of <u>enzymes</u>, the true catalysts of tissular chemical reactions: behind each "property" of life there are one or more enzymes. A decisive blow was thus dealt to vital forces once certain reactions, peculiar to living tissue, could be reproduced in the test tube.

With the biochemical study of the cell replacing cytology (which, with cell theory[4], had provided the first unifying theory of life), the concepts linked to energy supplying processes were posited. From then on, what characterised the study of life between the end of the 19th century and the middle of the 20th, was, above all, <u>cell energetics.</u> This "particular vision" of life, this "energetism", was part of the shift towards physicochemical aspects by biologists and also philosophers of the period who tried, henceforth, to define the properties and

---

3. F. Magendie, Elementary Handbook of Physiology (1825).
4. T. Schwann (1810-1882).

mechanisms of organisms as a set of <u>structures</u> and/or <u>reactions</u> using science which had been used to explain the inanimate world until then.

In this respect, Lavoisier (1743-1794) had been an inspired precursor with his work on oxygen and animal respiration which he compared to combustion. But between the end of the 19th century and the first decades of the 20th, Louis Pasteur (1822-1895) and the German and American enzymologists elucidated the main mechanisms responsible for sugar fermentation and tissue respiration (H. Krebs), while cell substructures, or " organels", (mitochondria, chloroplasts) were discovered at the same time relative to respiratory exchanges and/or photosynthesis in higher organisms. Biochemists thus emphasised two key processes: <u>oxidation-reduction,</u> which mobilises dehydrogenases and the cytochrome chain, true respiratory pigments (Keilin), and the key role of <u>phosphorylation</u>. ATP, a vital cell component the role of which is to act as the cell's energy centre was defined by the German Karl Lohmann (1898-1978). Closer to home, during the period following the Second World War, the American biochemist Fritz Lipman (1899-1986) (a refugee from Nazi Germany), introduced the concept of "energy-rich" substances, so called because hydrolysis of their covalent bonds with phosphorous radicals releases the energy necessary for the transphosphorylation of carbohydrates and other receptors. Phosphorous-containing metabolites can therefore be classified energetically, according to their "hydrolysis potential", which can be used to predict the evolution of a metabolic cascade leading to a succession of exchanges of phosphorous radicals within the cell.

However, although oxidation-reduction reactions and transphosphorylations involved in the breakdown of carbon structures provided a satisfactory explanation of how the cell draws its energy from the ambient environment through a multitude of enzymes and cofactors, the fact is that the cell is not simply a "micro power plant"! A tremendous problem continued to challenge biologists for decades: how the actual <u>formation</u> of cell constituents and their precursors takes place.

First of all let's discuss these precursors, which biochemists call "essential metabolites": amino acids (the constitutive elements of proteins), purine or pyrimidine bases (which enter into the composition of "nucleotides", themselves elementary constituents of nucleic acids), but also simple sugars and vitamins, etc. We often compare these metabolic precursors to "building blocks", the assembly of which results in the macromolecular structures present in the cell itself. These metabolites are provided in a preformed state or by digestive processes in animals. On the other hand, in plants and microbes, they are manufactured from more simple carbon compounds, such as organic acids, the chemical intermediaries of cell respiration, (Krebs' cycle), or C_3 or C_4 compounds , themselves derived from CO2 during photosynthesis. A bacterium, for example, can synthesize <u>all</u> of its cell constituents from a single and unique source of carbon such as glucose, ammonia salts, phosphates and mineral trace elements. So the chemical reactions responsible for the conversion of carbon compounds into essential metabolites are themselves complex. They often lead to a cascade of enzyme reactions, involving many chemical intermediaries, which are all the more difficult to characterise in that they do not accumulate in the cell. Isotope labelling with carbon 14 is often used to determine the origin of a particular carbon atom present in the chemical structure of the final metabolite, and the latter's contribution to the formation of macromolecules. But this technique has proved to be quite cumbersome and not very informative as regards the identification of reaction intermediaries. This is where genetics first became involved. In fact, applied to the study of biosynthetic stages in the production of the main metabolites in bacteria or some lower fungi (e.g. Neurospora), the isolation of "mutants", blocked at very precise stages of their biosynthetic process, led, within a few years, to a general description of the principal metabolic routes taken by the 20 amino acids which are the constituents of proteins or the five main nucleic bases[5].

---

5. Adenine, guanine, cytosine, thymine, uracil.

I.2. MOLECULAR BIOLOGY AND ITS ACHIEVEMENTS

I.2.1. MOLECULAR BIOLOGY OF THE GENE (DOUBLE HELIX, GENE EXPRESSION AND REGULATION, THE "CENTRAL DOGMA")

Tackling protein and nucleic acid synthesis required a completely new concept to be devised, as well as a methodology and original techniques. A link had to be made between disciplines which had not previously been related, such as biochemistry, genetics and macromolecular physicochemistry, often by using theories relating to other fields of knowledge, such as information theory.

The new "paradigm" (this expression was coined by the science historian, R. Kuhn) to which this set of questions, concepts and techniques subscribes, was known as molecular biology. Just as the discovery of enzymes and energy metabolism fuelled the first unifying concepts in the life sciences, from the early 1950s the idea of the gene as a basic unit of heredity changed the entire construction of contemporary biology.

• *DNA and the double helix*
In 1946, the physicist Erwin Schrödinger stated in his book *"What is Life ?"* that the foundations and principles of life's functions are to be found in the properties of large polymers, or biological macromolecules, proteins and nucleic acids. As regards proteins, biologists had very little difficulty in agreeing to this idea. Enzymes are proteins, antibodies are proteins, proteins are the major constituents of cells, etc. Their role even seemed so central in cell life that it took real courage for the American pathologist R. T. Avery to assert, with experiments to back it up, that genes are not proteins but nucleic acids, and that hereditary material is made of one of them, deoxyribonucleic

acid (DNA), a phosphorous biopolymer isolated from salmon sperm in around 1860 by the Basle native F. Miesher, (and which was considered, for decades, to be a somewhat mysterious, if not incongruous, cell constituent with respect to function). It is to Schrödinger's credit that not only did he determine the major role of nucleic acids in cell life but also did the groundwork which formed the basis of a new structural biology in showing that the key to the most important cellular functions lay in the physicochemical structure, properties and interactions of macromolecules.

One of the most spectacular illustrations of this molecular reductionism, and the most well-known, was undoubtedly the discovery of the DNA double helix structure in 1953 by J. D. Watson, F. Crick and M. H. F. Wilkins. The fact that the helix is double, and that each of its strands is a perfect match for the other in accordance with the laws of perfect complementarity (A-T; G-C) through weak hydrogen bonds, would provide key explanations for hereditary transmission in dividing cells. In their original publication, furthermore, Crick and Watson already anticipated how this DNA "replication" took place, with the separation of each strand, their distribution within daughter cells, followed by the reconstitution of a double helix according to the laws of ATCG base pairing previously mentioned, foreshadowing the existence of enzymes responsible for this reassembly. The first DNA polymerase was discovered a few years later (A. Kornberg...) and although time and experimentation would show that DNA replication involves much more complex enzymatic mechanisms *in vivo* (topoisomerases, DNA ligases, etc.), Komberg's discovery demonstrated, for the first time, that a process of macromolecular biosynthesis can occur "outside" the cell. Attempting to reproduce some of the key reactions of cellular chemistry *in vitro* then became one of the preferred approaches of molecular biology.

Fig. 1 *The DNA double helix*

A/ It was in 1953 that J. D. Watson and F. H. Crick described the crystalline structure of the DNA double helix for the first time [Nature, 171, 737, (1953)]. The double helix quickly became symbolic of molecular biology. M. Willeins shaved the Nobel Prize and R. Franklin played an important role as well.

B/ Replication of the double helix: note the separation of the two complementary strands and the reconstitution of the two daughter helices. In the cell, this process involves an enzyme complex (replicases, helicases, etc.) and factors for destabilising the double strand structure.

C/ Principle for pairing base pairs AT and GC, from Molecular Biology of the Gene [(1965) p. 267, fig. 9-7 and p. 132, fig. 4-14 – reproduced in "Les secrets du gène", François Gros, Éditions Odile Jacob (1986) p. 63]. The crucial roles played by Maurice Wilkins (Nobel prize, 1962 with Watson and Crick) and Rosalind Franklin (died 1958) must not be forgotten.

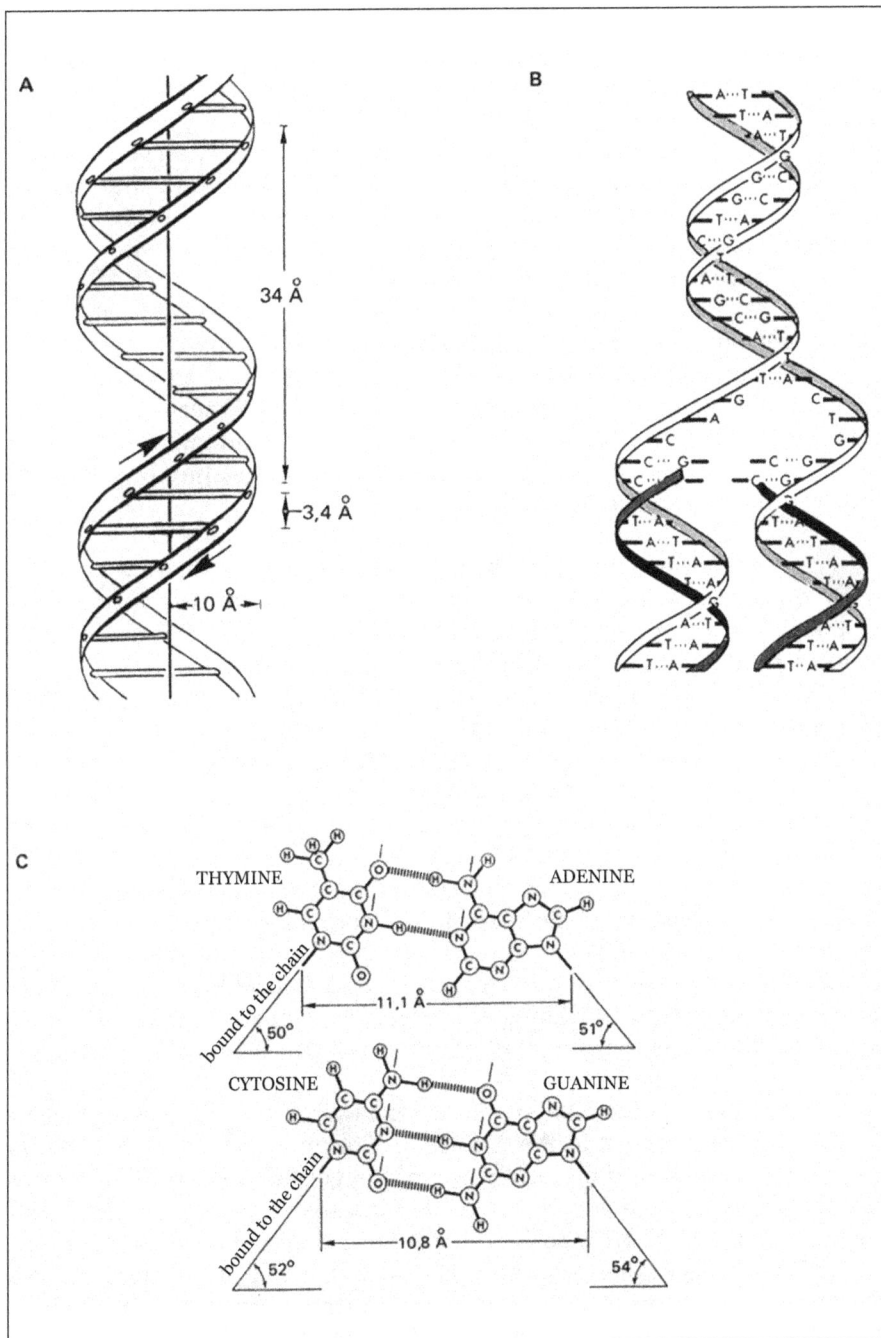

A

34 Å

3,4 Å

10 Å

B

A···T
T···A
A···T
G···C
G···C
T···A
C···G
A···T
G···C
C···G
A···T
T···A
G   C
A   T
C   G
C···G   C···G
T···A   T···A
A···T   A···T
T···A   T···A
T···A   T···A
A···T   A···T
T···A   T···A
T···A   T···A

C

THYMINE   ADENINE

bound to the chain
50°   11,1 Å   51°

CYTOSINE   GUANINE

bound to the chain
52°   10,8 Å   54°

Although the characterisation of DNA structure was the first major contribution of molecular biology, assisted by the work of crystallographers, the study of proteins also benefited from physicochemical techniques often linked with crystallographic analysis. The work of L. Pauling (the discovery of the α helix structure) was also followed, a few years later, by crystallographic studies by M. Perutz and J. Kendrew, which led to the first three-dimensional models of proteins (fibrous and globular) shedding entirely new light on the physicochemical representation of enzyme active sites, allostery (Monod, Wyman, Changeux) and antigen-antibody interactions.

• *Gene function and regulation (initial concepts)*
The great mystery which molecular biology began to investigate from the end of the 1950s was gene function, a broad outline of which had been laid down several years previously by analysing the effects of mutations on certain characteristics of the fruit fly (the famous Drosophila), the fungi *Neurospora* and, later on, bacteria and bacteriophages.

Already towards the end of the 1940s some American geneticists (Beadle and Tatum), and in France B. Ephrussi and P. Lheritier were wondering about gene action on a molecular level, thus creating a new branch of genetics which they called "physiological genetics". In a startling short-cut, Beadle and Tatum formulated their famous equation, memorable to scientists of the period: "*one gene, one enzyme*". Although this idea has now been shown to be inaccurate (one gene may determine several proteins; some genes code for nucleic acids, etc.) it still had the huge merit of highlighting the real problem: genes, usually govern protein manufacture in cells, and this provides the most simple basic explanation of the effect of mutations, in the knowledge that many proteins are enzymes involved in metabolism, the alteration of which can result in the modification of a characteristic linked to the morphology, growth and survival of a living organism. A link between classical genetics (G. Mendel, T. Morgan) and biochemistry had therefore been made. Biology could now tackle "gene-protein" relationships by stating the problem in precise molecular terms corresponding to the transfer of information between DNA and proteins.

This new molecular biology of the gene proved to be extraordinarily fruitful. We are indebted to this branch of biology for having elucidated, in a very short period of time, the nature of the genetic code and the various biochemical stages intervening in the two successive "readings" of the "code" inscribed in DNA: the transposition of the DNA chemical code to ribonucleic acid (RNA) – a stage called transcription – then, from this "RNA intermediary", the formation of a protein, or translation. Furthermore, the work of biologists at the Pasteur Institute, particularly F. Jacob and J. Monod, introduced a concept of considerable significance in our understanding of life, by demonstrating that the thousands of genes present in a cell (for example in a bacterium) are transcribed and translated according to a defined programme, reflecting not only the necessary coordination and intricacy of innumerable metabolic and cell development reactions, but also the physiological adaptation of the cell to its environment. In other words, the concept of cell regulation, which could be "positive" (gene activation) or "negative" (repression), depending on external environmental influences and also (in higher organisms) the level of tissue development, was proposed for the first time.

Before returning briefly to these great breakthroughs in molecular biology applied to life studies, mainly between the early 1960s and the early 1980s, a preliminary comment must be made. In most cases, it was the use of biological models of bacteria and their viruses (bacteriophages) which clarified the key stages in gene transfer and regulation. So, when biologists turned to the study of these phenomena in higher organisms, they discovered that, while many of the rules concerning the mechanisms of gene expression as derived from the study of bacterial models were globally applicable to eukaryotes, the mechanisms in the latter were infinitely more complex. It was particularly shown that certain characteristics, both structural (split genes, regulatory DNA sequences, DNA configuration in chromosomes) and functional (splicing, interference phenomena, post-transcriptional regulation, reverse transcription, etc.) were exclusive to evolved organisms, called eukaryotes, apart from rare exceptions[1].

I.2.2. THE GENETIC CODE – THE TRANSFER OF GENETIC INFORMATION: TRANSCRIPTION AND TRANSLATION

As we have already emphasised, the main challenge which molecular biologists faced in the middle of the 1960s, was explaining the "transfer" of "information" from DNA to proteins, through the two essential steps of DNA transcription to RNA and translation, which converts the information thus transposed to this RNA intermediary, later designated messenger RNA (see below), resulting in the final product which constitutes the protein. It was realised quite early on, however, that it would be advisable to understand exactly what the 'information' consists of in nucleic acids! Solving this problem of the genetic code would solve one of the great "mysteries of life" as many biologists called them, semi-ironically, semi-seriously.

---

1. See especially chapter I-4 entitled "The complexity of genetic material in higher organisms".

The physician G. A. Gamow (1904-1968) first understood that genetic information resides in the <u>chemical sequence</u>, i.e. in the order of <u>assembly of the basic units which</u>, like letters in a sentence, are linked to each other to make up each strand of the double helix. The biologists Francis Crick and Sydney Brenner completed the proposition brilliantly, demonstrating that everything comes down to understanding how a chain of four "motifs" – the four A, T G and C bases aligned in the DNA molecule – can <u>code a</u> chain of twenty protein "motifs", namely the twenty types of amino acids present in the series of known proteins. The calculation demonstrated that binary combinations of nucleic bases would be too few (4^2=16), while ternary combinations (4^3=64) satisfy the prerequisites by assuming that there are, on average, three ternary combinations, or "triplets" per amino acid. Crick and Brenner, in very refined experiments, confirmed these predictions experimentally: the genetic code is therefore a "triplet" code. The cell "reads" the DNA sequence, starting at a specific end of the chain, but with no "overlapping" and with great care so that there is no ambiguity in the mode of linking the amino acids.

Fig. 2 *Main stages in gene expression (for a prokaryote cell)*

A/ One of the two strands of the DNA double helix is copied by an enzyme DNA-dependent (RNA polymerase) onto a continuous chain of RNA. This operation begins at a special sequence known as the "promoter" and ends at a termination site. Before the RNA chain is detached, a ribosome is bound to a start site (e.g. AUG) and the polypeptide chain is elongated. Movement of the first bound ribosomes towards the messenger RNA terminal sequence (UAA in this example) allows other ribosomes to be bound. The dynamic structure of the "processed" RNA is shown by a string of ribosomes known as a "polyribosome" (A. Rich).

In a prokaryotic cell a ribosome is a complex organelle formed of 2 separable subunits. The polypeptide takes on a three-dimensional conformation before the end of its elongation, after which it is detached and takes on its final configuration.

B/ The entire process described in A can be reduced to 2 main stages:

i) Transcription: or copying DNA (double strand) onto RNA (single strand).

ii) Translation: an operation during which messenger RNA acts as a template for protein formation.

The situation described here applies to genes coding for proteins. Some genes are copied onto non-messenger RNAs. These are not translated but can perform various functions in the cell (rRNA, tRNA, siRNA, miRNA, etc. ; for further details see the chapter "The world of RNA").

THE MAIN STAGES
OF GENE EXPRESSION

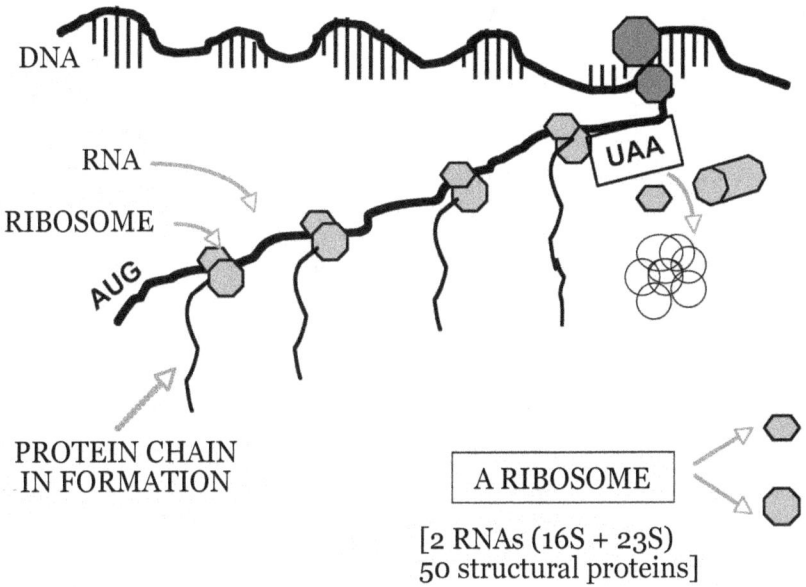

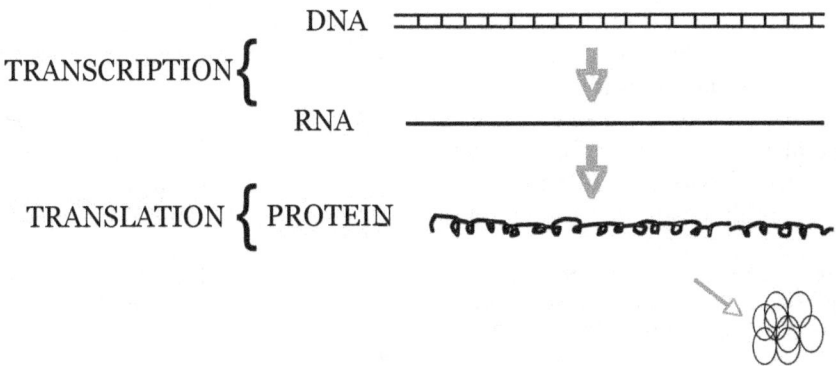

DNA

RNA

RIBOSOME

UAA

AUG

PROTEIN CHAIN
IN FORMATION

A RIBOSOME

[2 RNAs (16S + 23S)
50 structural proteins]

TRANSCRIPTION {

DNA

RNA

TRANSLATION { PROTEIN

Incidentally, we can see that the code is not one-to-one (there is more than a single triplet on average per amino acid). In cryptographic language the code is said to be "degenerate", but while the principles of the genetic code were clearly established in this way, it still remained to clarify its chemical nature, i.e. to specify which triplets, out of 64 possible sets, correspond to which amino-acids. Another question had still to be answered: what exactly is "the RNA intermediary" which has a sequence reflecting that of DNA and acts as a template for the assembly of protein chains.

• *Messenger RNA and the genetic code*
     In 1961, two groups of researchers, one working in Pasadena (S. Brenner, F. Jacob and M. Meselson) and the other in Harvard (F. Gros, J. D. Watson and coll.), established the nature of the intermediary or, if you prefer, the messenger RNA acting as a template for the assembly of polypeptide chains. In bacteria and bacteriophages (where the "messenger" was identified for the first time), it is a chain of ribonucleic acids with a composition of bases and a sequence which accurately reflects the composition of one of the two DNA strands[2]. Shortly after the discovery of messenger RNA, produced by the primary transcription of DNA, the enzymes responsible for this transcription were discovered, RNA polymerases, also called "transcriptases" (S. Weiss, G. Hurwitz).

     The chemical nature of the genetic code was determined in record time, through the use of artificial polymers, types of polyribonucleotides containing only a single base, (or combinations of more, experimentally determined, bases). As had previously been shown by the biochemists M. Grunberg-Manago and S. Ochoa, biosynthesis of these polymers results from the activity of the the enzyme polynucleotide phosphorylase. The nature of the genetic code was then deciphered following a series of experiments implementing these artificial templates (M. Nierenberg and Mathaei, S. Ochoa and A. Wabba). It emerged that 61 of the 64 possible triplets were "significant", i.e. they correspond to an amino acid specific to each of them; as for the three remaining triplets, their role was established as " punctuation" factors, that is, acting as chemical signals, either for initiation of the assembly of polypeptide chains (in the case of AUG, or GUG triplets), or as stop signals and for the release of the polypeptide chain on termination of the assembly (UAA and UAG triplets).

---

2. Messenger RNA is therefore capable of experimentally forming a hybrid heteropolymer, DNA: RNA, in which the bases of each of the two chains are completely paired (B. Hall and S. Spiegelman); its average lifespan in the cell is short, lasting around several minutes. Messenger RNA, molecular imprint of DNA, is therefore the genuine information template on which the amino acids are aligned in turn through the formation of covalent bonds (peptide bonds) generating the long polypeptide chain which forms the backbone of the protein.

It was later shown that the genetic code is universal: the same chemical combinations are used in all living things, from bacteria to man.

The discovery of the genetic code and its "universality" at the start of the 1960s is therefore quite rightly considered to be not only one of the major accomplishments of molecular biology, but also one of the greatest scientific triumphs of the 20th century. The code's universality, in a way, prefigured the near-explosive proliferation of transgenetic experiments following the development of genetic engineering which will be discussed a little later on.

• *Protein synthesis*
Having examined how the amino acid assembly code was discovered, we shall now return to the important stage in cell function known as "genetic translation". It is an important stage because it covers all the mechanisms as well as the "devices" responsible for protein formation. To understand the procedure, biologists had to use both theoretical and empirical methods. Protein synthesis requires a template alignment of amino acids which can code their successive "positions" opposite the codons in messenger RNA before their chemical binding by peptide bonds. This template is none other than messenger RNA (see above). However, elongation, a kind of progressive manufacture of the protein chain in contact with the RNA requires the equivalent of a "pickup". This is provided by a complex organelle, the ribosome. Its role is to gradually move along the strand of messenger RNA like a "rack railway", placing the corresponding amino acid opposite each nucleotidic triplet (each codon), according to the rules of the genetic code. A polypeptide chain is thus gradually produced (the elongation stage). On reaching a codon stop (UAA, UGA), the ribosome detaches itself from the messenger RNA and releases the completed polypeptide chain attached to it. This spontaneously folds into the three-dimensional conformation of the protein in question (C. Anfinsen).

This process of genetic translation, therefore starts, operationally, at a specific site in the messenger RNA chain called the "start codon" (the AUG or GUG codon). Specific protein factors, called initiation factors (M. Revel and F. Gros) ensure "recognition" of this particular site by the ribosome, which marks the start of elongation of the protein chain. So RNA-protein translation is limited by the start and terminal codons. The whole process requires energy which is generated by hydrolysis of a particular related component of ATP, guanosine triphosphate, or GTP. But reading the messenger RNA and translating it into protein would hardly be possible if each amino acid had to be bound "directly" opposite the corresponding codon. Amino acids display different molecular structures: the radicals which make the difference between them do not occupy the same dimensions in space and their resulting steric size would hinder favourable alignment with the regular manufacture of the growing polypeptide chain.

In the mid-1950s, the solution to this problem was provided through the work of molecular biologists P. Zamecick, M. Hoagland and, in particular, Francis Crick.

In short, the sequential "positioning" of each amino acid opposite its codon, a procedure which we have seen take place at the ribosome-codon interface, takes place via a specific "adaptor" which carries the amino acid at its end. It is a small RNA molecule, called "transfer RNA" or t-RNA, whose shape resembles a clover leaf. One of the ends of the adaptor RNA – in reality a short sequence called "anti-codon" – forms complementary bonds by base-pairing with the messenger RNA codon to "match" it, while the other end holds the amino acid that specifies the "messenger" codon. The amino acid is bound to its transfer RNA by an enzyme, aminoacyl RNA synthetase (also called an activating enzyme) specific to both the amino acid and its adaptor. So, for each of the 20 amino acids there is a tRNA attachment enzyme, the tRNA being itself specific and able to recognise the codon for that amino acid.

The extraordinary precision and great complexity of genetic translation systems, which function with a very low margin for error, raises interesting problems in terms of molecular evolution. We are led to wonder how a functional assembly so closely adjusted could have been the subject of natural selection. What type of coevolution has taken place to ensure that the genes coding for transfer RNA and aminoacyl RNA synthetase enzymes could have acquired specificity for the same amino-acid? A number of prebiotic chemists have also suggested that, originally, there were undoubtedly mechanisms for "direct" recognition between RNA and amino-acids.

I.2.3. GENE REGULATION – THE REPRESSOR – THE LACTOSE OPERON

This global picture describing the major functions of the cell, which was drawn by scientists during the first phase of molecular biology, lacks a major perspective, as was already pointed out. Indeed, all behaviour, as depicted by cell physiology demonstrates a high degree of plasticity. A bacterial cell will not manufacture the same enzymes, depending on the composition of its culture medium; a cell within a tissue or differentiated organ from an animal or plant will preferentially synthesize and accumulate certain proteins specific to the tissue or organ studied and this may vary markedly during development. In other words, there must be <u>regulatory mechanisms</u> which modulate, prevent or activate certain genes according to the conditions imposed by the external environment (bacteria) or through tissue differentiation (eukaryotes).

At the start of the 1960s J. Monod and F. Jacob proposed a model of <u>negative regulatory mechanisms and the operon</u>, revealing, for the first time, the existence of a genetic regulatory loop in a living organism, the bacterium *E. coli*.

After a logical series of observations relating to the growth of this bacteria in the presence of two sources of carbon (glucose and lactose), research by these scientists from the Pasteur Institute revealed the process of <u>adaptation,</u> using lactose to demonstrate that this metabolic adaptation is based on the neosynthesis of a metabolic enzyme, β-galactosidase. Following successive studies, Monod, Jacob and their colleagues developed their famous model, usually referred to as "induction of the lactose operon". It was established that lactose, or a chemical analogue of this sugar, acts as an " inducer" which can <u>overcome the negative effect of a "repressor",</u> <u>blocking the transcriptional activity of the 3 genes responsible, in *E. coli*, for</u> <u>the metabolism of lactose or similar sugars</u>. In other words, French researchers succeeded in demonstrating that, in the majority of cases, at least in bacteria , <u>many genes responsible for general metabolism remain "silent"</u> <u>in the absence of environmental factors acting as inducers</u>. The study of the "lactose system" was gradually completed through a series of genetic experiments. The repressor was isolated a few years later by W. Gilbert and B. Müller Hill. It is a protein that can inhibit transcription of the three key genes involved in β-galactoside metabolism: β-galactosidase, galactoside permease and galactoside acetylase. These three genes are contiguous on the chromosome and during their unidirectional transcription (after induction) at a specific "promoter" site. If the inducer is absent, the repressor attaches itself, with great affinity, to a site next to the promoter, called the "<u>operator</u>". It prevents RNA-polymerase from binding to this promoter, thus inhibiting the unidirectional transcription of the three adjacent genes on the chromosome. The name <u>operon</u> was given to the set of genes responsible for a given metabolic function when they are topologically adjacent. So we talk about the "lactose" operon. In the presence of an inducer (lactose or some of its chemical analogues), the repressor undergoes an allosteric transformation and detaches itself from the operator leaving the field open for the polymerase responsible for cotranscription of the operon.

In the general presentation of their work, in 1960, Monod and Jacob postulated the existence of two categories of genes: "<u>regulatory genes</u>", coding for agents exerting negative control over "classic" genes, also called "structural genes" because they determine the synthesis of enzymes or proteins endowed with other functions. The discovery of these two categories of gene and their presence in all living organisms, including eukaryotes, therefore opened a fundamental and entirely new chapter of biology. However, whereas at the start of their work Monod and Jacob thought that the control exercised by regulator genes was always "negative" (repression-derepression), it was later shown that a number of genes belonging to bacterial operons (e.g. the "arabinose" or maltose operon) exercise "positive" regulation. This means that their cotranscription depends on an activating protein factor, often an exogenous metabolite. Subsequently, when biologists were able to study the genes of higher organisms, it was found that their activity also usually responds to positive control. What is more, these eukaryotic genes are never, or hardly ever, organised "in tandems" or operons, contrary to what is seen in bacteria (see below). It is nonetheless true that concepts of regulatory genes and structural genes are fully applicable to regulation processes in " eukaryotes" and that there is a general functional hierarchy in genetic material.

I.2.4. THE CENTRAL DOGMA OF MOLECULAR BIOLOGY

So we can say that, towards the mid-1970s, the molecular approach to biology had led to an immense wealth of discoveries! Actually, the main reproductive and biosynthetic functions of the cell had been explained. Biologists could entertain the illusion that the main cellular mysteries had been solved. Two aphorisms seem to describe this apparent certainty very well. The first, semi-ironic, semi-serious, is based on the idea that all of biology boils down to a kind of central dogma, now revealed, a dogma which seems unshakeable and is expressed in the form of a concise equation: "DNA→RNA→protein" implying that the flow of information is unidirectional. The second, of a very different order certainly, but not least revealing of a somewhat triumphal sentiment, concerns the universality and exhaustiveness of biology discoveries. According to this idea, beloved of a number of molecular biologists of the period, especially the great scientist Jacques Monod: "what is true for *E. coli* is true for the elephant".

Although it is still true in general, this famous "central dogma" of molecular biology was found to have a major exception when, around 1975, Baltimore and Termin, studying the reproductive cycle of certain retroviruses, observed, for the first time, that genetic information can also flow from RNA to DNA, under the effect of "reverse transcriptase". It was later realised that this retro-transcription is also observed in the cell environment itself (retrotransposons). Some biologists claim that the ability of certain living systems to copy an RNA sequence into a DNA sequence probably appeared very early on, in the prebiotic era, making <u>RNA (and not DNA) the first informational molecule.</u> As for the idea that all the important things had been explained through the molecular biology of *E. coli*, the future, even here, proved this to be premature, when genetic engineering made the nature and functions of genes in higher organisms accessible to researchers.

I.3. GENETIC ENGINEERING – BASIC CONSEQUENCES – APPLICATIONS

I.3.1. GENETIC ENGINEERING – DISCOVERY – BIOLOGY OF HIGHER ORGANISMS

The discovery, development and applications of genetic engineering represent a major epistemological rupture in the history of modern biology for many reasons. The most important, on a basic level, is that this technology introduced by Berg, Chang and Boyer at the start of the 70s, led to the purification and systematic isolation of genes from higher organisms, which had been practically impossible up until then. To describe things very briefly, this technique first involves cutting the DNA isolated from the eukaryote cell using specific nucleases known as restriction enzymes (W. Arber), which splits the DNA molecule into extremely precise sequences. The fragments obtained are then "recombined" *in vitro*, using binding enzymes called "ligases", with "vector" DNA. This usually consists of bacterial plasmids, small circular DNA sequences which can transfer (transfection) into bacterial host cells by inserting themselves into the host DNA. The experimentalist ensures that each receptor bacterium only integrates, on average, one "plasmid – recombinant DNA fragment". Each bacterium, transfected in this way, is cultured on a solid medium. After several divisions, a colony (clone) is formed of bacteria which are absolutely identical to each other and to the original bacteria, and contain the same eukaryotic DNA fragment. The fragment amplified in this way in each of the clones is then pinpointed, using an *in situ* hybridization procedure by means of an appropriate molecular probe (generally a radioactively-labelled complementary DNA). The bacteria which constitute this clone are then cultivated on a large scale and the total DNA extracted. This contains the particular eukaryote gene of interest embedded in its sequence. This gene is then detached from the surrounding plasmid DNA sequences using restriction enzymes.

This operation is generally described as "cloning". In the jargon used by biologists, we talk, for example, about "cloning the albumin gene", or in short, "albumin cloning". Cloning guarantees, (by definition), total purification of practically any gene, whether of prokaryotic (bacteria, inferior algae) or eukaryotic (animals, plants, fungi) origin. We will see shortly what the resulting theoretical consequences are with respect to the molecular organization of genes in higher organisms.

Another profound upheaval, conceptual as well as methodological, which followed the rapid development of genetic engineering is the possibility of transfering a gene purified by cloning and originating in a species or a defined kingdom, to cells or whole organisms of a different species or biological kingdom. This procedure was dubbed "transgenesis"; it produces genetically modified organisms, the notorious GMOs, used on a vast scale in various countries and more or less considered with reservation in others (see the chapter devoted to GMOs). With transgenesis, genetics has considerably increased its power of intervention on living organisms, in medicine as well as agriculture, thereby generally reviving biotechnology .

Whereas molecular biology of the gene developed mostly using the bacteria *E. coli* as a model microorganism, had presented an image of a remarkably logical series of achievements and concepts, mostly developed on the basis of information theory and cybernetics, the study of genes in higher organisms revealed completely unpredictable distinctive features! The "logic of life" (the term is François Jacob's) has many surprises in store for us.

We can be sure that very few molecular biologists in the first wave would have imagined that "eukaryote" genes would be formed differently from those of bacteria, i.e. that they would be represented by a mosaic of discontinuous sequences or segments inside the long DNA chain present in the chromosomes. This description seems all the less plausible in that certain biologists (Dintzis, S. Brenner) had revealed the existence of a precise co-linearity between DNA and protein ... a one-off mutation in DNA which changes a single amino-acid at a specific point in the corresponding protein chain.

I.3.2. EXONS-INTRONS

So there was great surprise when the American researcher and R.J. Roberts who shared the 1993 Nobel Prize P. Sharps, and then biologists such as P. Chambon and Ph. Kourilsky, showed that genes in animal cells, for example, are "mosaics". In most cases they are, in fact, comprised of exons, "coding" segments, which link introns, non-coding DNA segments. This "mosaic" layout is all the more astonishing in that the length of the introns can vary considerably and the rules governing the location and number of introns are still not fully known. Bacterial genes do not contain introns, the genes of Archaebacteria, organisms belonging to an intermediary evolutionary lineage between "classical" bacteria (eubacteria) and eukaryotes, sometimes do, as do eukaryotes. The predominating hypothesis today is that the bacterial cell, which appeared very early on during evolution (2.5 to 3 billion years ago), most probably contained introns in its genome but they would have been eliminated during the innumerable divisions which accompanied their evolution.

Moreover, researchers such as Walter Gilbert have provided a convincing theory to explain the existence of mosaic genes. The first coding elements which appeared during evolution must have been very short DNA sequences presumably encoding small peptides, with limited catalytic activities. Through the gradual assembly of these "modules", longer polypeptides could have formed, then proteins endowed with a three-dimensional structure compatible with various enzymatic functions.

Faced with the discovery of mosaic genes, biologists were confronted with the following question: How could a gene sequence with <u>discontinuous</u> coding modules have been transposed into a <u>continuous</u> messenger RNA sequence, with a sequence co-linear to that of the corresponding protein?

I.3.3. Splicing

The phenomenon which explained this dilemma is called "splicing". Firstly the discontinuous nuclear genes are transcribed over their entire length. The RNA copy thus obtained therefore includes complementary sequences of exons linked to complementary sequences of introns. The mosaic transcript is then transported from the nucleus to the cytoplasm and the RNA sequences corresponding to the introns are eliminated when the transcribed portions derived from single exons are inserted. This process is catalysed by rather complex cell organelles, called <u>spliceosomes</u>, themselves comprised of a series of proteins and specific RNAs. The net result is the formation of a chain of messenger RNAs from a mosaic-gene leading to a situation in which the messenger RNA information and the protein are colinear!

The discovery of this singular mechanism <u>led to two major observations</u>:

Firstly, there appeared to be a post-transcriptional regulation mechanism widely used in eukaryotes known as alternative splicing. In fact, depending on the stage of tissue development and sometimes on the cell signalling factors involved, the **nature** of the RNA sequences transcribed from exons and spliced can vary. **The result is that through this selective connecting mechanism, the same gene can code for several proteins of different compositions or, more simply, several proteins can correspond to one gene.** When a gene comprises a large number of exons, the combinations resulting from alternative splicing can be very diverse: some genes can code individually for hundreds of proteins! This "molecular tinkering" represents a considerable evolutionary advantage for higher organisms. It actually provides the eukaryotic cell with an infinitely greater variety of proteins, more varied that that observed in a bacterium such as *Escherichia coli*. This situation will be discussed further in our section on proteomics. It is also a mechanism of major importance in cell differentiation.

But another general observation, which was even more unexpected, led to the discovery of the ribozymes.

I.3.4. Ribozymes and the world of RNAs

Some "primary transcripts" corresponding to the "exons-introns" series can, in the absence of spliceosomes and protein catalysts, spontaneously excise some RNA sequences, copies of some of their introns. This case of self-splicing was described for the first time in the mid- 1980s by T. Cech. Since

Fig. 3 *Simplified diagram of the stages of gene expression in a eukaryote*
You can see a gene consisting of coding regions of DNA known as "exons", linked by "separators" (usually non-coding) known as "introns". (Here the gene includes 4 exons and 3 introns).
– The first stage, called transcription involves enzymatic copying (RNA polymerase) of the gene into pre-mRNA (pre-messenger RNA), a partial replica of the "exon-intron" unit.
– The next stage, called "splicing" involves joining RNA copies of exons, eliminating RNA copies of introns and the formation of mature messenger RNA.
Note that splicing is often "alternative" as shown here. RNA copies of exons can be assembled in different combinations (here: 3 combinations of messenger RNA: exons 1-2-4, 1-3-4 and 1-2-3-4).
The mature messengers produced by splicing are exported from the nucleus to the cytoplasm, where they are "translated", producing 3 different proteins.
(reproduced from Biofutur, n° 216, p. 40, 2001.)

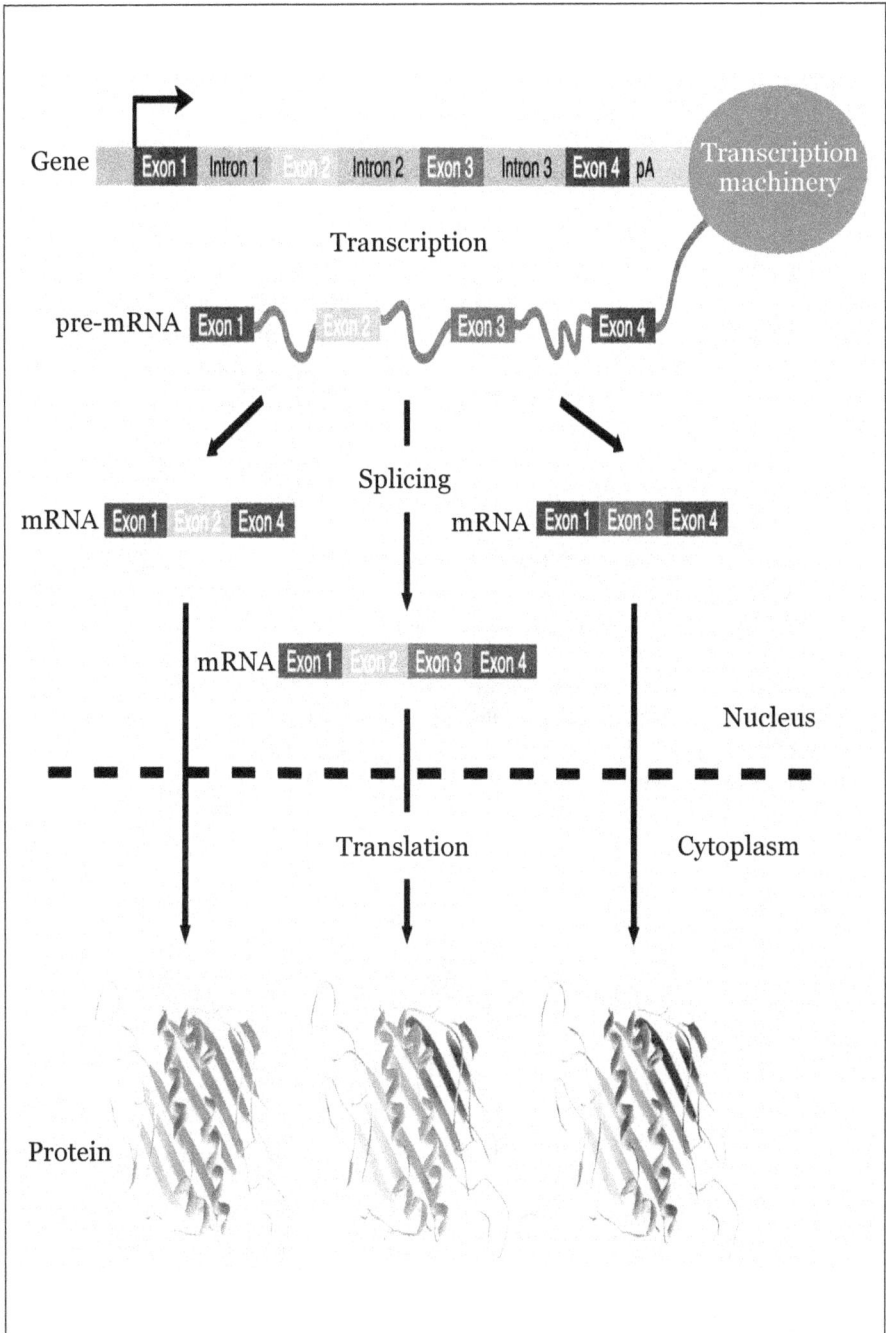

Gene   Exon 1 Intron 1 Exon 2 Intron 2 Exon 3 Intron 3 Exon 4 pA   Transcription machinery

Transcription

pre-mRNA Exon 1 Exon 2 Exon 3 Exon 4

Splicing

mRNA Exon 1 Exon 2 Exon 4   mRNA Exon 1 Exon 3 Exon 4

mRNA Exon 1 Exon 2 Exon 3 Exon 4

Nucleus

Translation   Cytoplasm

Protein

then, other catalytic activities pertaining to some RNAs have been demonstrated, the sole prerequisite being the presence of divalent cations. The name
ribozymes was given to these RNAs as a reminder that they behave like real
enzymes (endowed in this case with nuclease activity). A number of them not
only show self-cleaving ability but may also attack foreign RNA molecules, for
example those of viral origin.

The discovery of catalytic activities displayed by some RNAs has over-
turned a number of concepts about the origins of life in the prebiotic era. The
current prevailing opinion is that **the world of RNAs** would have played a
dominant role among the very first macromolecules that appeared. The existence of reverse transcription systems (RNA → DNA), although widespread
today in certain viruses, may lead us to suppose that DNA would only have
appeared afterwards. More generally, we shall see that the world of RNAs still
holds some major surprises concerning the astonishing, recently discovered,
phenomenon of "interference".

The discontinuous nature of the genes which form an integral part of its
sequence is not the only property which distinguishes eukaryotic DNA from
that of bacterial prokaryotes or bacteriophages.

In fact, another essential difference lies in the respective cytological
locations of these two classes of DNA. Eukaryote DNA exists in an extremely
compact form in chromosomes inside a cell nucleus, while that of prokaryotes
is more or less in direct contact with the nucleus-free cytoplasm.

I.4. THE COMPLEXITY OF GENETIC MATERIAL IN "EUKARYOTES"

I.4.1. CHROMATIN COMPACTION – NUCLEOSOMES

The compaction of eukaryotic DNA has been the subject of a great deal of work. It must be understood that an animal or plant cell a few microns in diameter contains nearly two metres of coded DNA ribbon inside its nucleus! This DNA is compacted through a succession of coiling and supercoiling... Firstly, the DNA double helix, which has some flexibility in this respect, is neatly coiled around minuscule protein corpuscles, resembling more or less spherical bobbins, the <u>nucleosomes</u>, by forming two loops around each of them, so that the set is presented as a "string of beads". Each nucleosome is comprised of a conglomerate of very basic proteins, characteristic of eukaryote cells: the <u>histones</u>. Further interactions cause this first type of coiling to undergo twists which give it the appearance of a "solenoid" which is "supercoiled" inside the chromosomes, thanks to the intervention of non-basic proteins, distinct from histones.

In bacteria (Rouvière-Yaniv), although DNA does not undergo such a complex system of coiling, and there is no indication of nucleosomes or histones, the DNA is nevertheless associated with proteins with high isoelectrical points, such as the thermostable Hu protein. The complex thus formed is similar to a "chromoid". The DNA is circular and attached to the membrane (F. Jacob, Ryter).

I.4.2. EPIGENETIC MODIFICATIONS

DNA transcription and replication in a eukaryotic cell necessarily involve major topographical rearrangements of the chromatin, leading to dissociation of the nucleosomes during the passage of polymerases followed by their regeneration. Some areas of chromatin will remain "compacted" and non-functional during development (heterochromatin). On the other hand, in the "active" areas, i.e. those which are transcribed (areas which can be located experimentally, due to their sensitivity to Dnase-1) a relaxation (provisional dissociation) of the nucleosomes, as we have just said, occurs. This follows acetylation of certain constitutive histones because the acetylated histones in lysine residues have very little affinity for DNA. The role of histone-acetylates is therefore crucial in the activation of certain portions of chromatin, while histone-deacetylates produce the reverse effect. Another way of locking eukaryotic genome transcription consists of methylating some of the DNA sequences near the promoters, such as CpG sequences. DNA-Methylase actually converts the cytosine in these CpG "islets" into methylcytosine, which obstructs nearby gene function. The name **epigenetic modification** is often given to these general modifications which are more or less reversible, and hereditarily transmittable, but do not alter the DNA sequence, these modifications particularly including histone acetylation and deacetylation and methylation (or demethylation) of the CpG islets.

We shall return in more detail to the mechanisms involved in this epigenetic regulation, which has an increasingly important role in eurkaryote development and dysfunction.

I.4.3. POSITIVE REGULATION – PROMOTERS – CIS-REGULATORY SEQUENCES

Another particularity of eukaryote DNA which merits attention is its wealth of regulatory sequences and repetitive elements. We have seen previously that prokaryote genes carry a start site for transcription to messenger RNA, a particular sequence called "promoter", where the RNA polymerase is attached. This process may be blocked by binding a repressor to an adjoining site called "operator", as is the case in negative control of the lactose operon. In other cases it may be activated both by removing repression and by the positive action of certain specific factors which are also set prior to the promoter (which entails the participation of a particular cofactor, cyclic AMP).

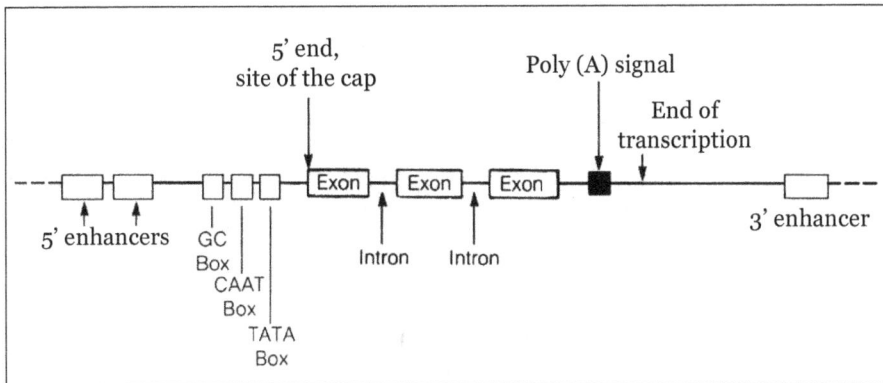

Fig. 4 *Regulatory sequences of a higher organism (eukaryote) gene*
(For the meaning of the symbols, see the text.) You can see various types of "punctuations" (cis-regulatory sequences), some of which are ubiquitous (CAAT, TATA, GC). Others, not shown here, can be specific: e.g. the sequence CNNTG (where N can be any base) usually accompanies gene sequences coding for muscle functions. "Enhancers" (activators) are also specific sequences. They can be located at the 5' or 3' end of the gene (gene transcription starts close to the TATA box and moves from the 5' end towards the 3' end). Most eukaryotic gene transcripts end with a poly A sequence (role in stabilising and transport of the primary transcript).

In eukaryotes as in prokaryotes, we see regulatory mechanisms involving promoters and regulatory sequences. Evolution has, however, given these organisms much more varied regulatory elements: for example, gene transcription usually begins close to a short sequence, TATA – known as "TATA Box". This process mobilises various factors, the role of which is to facilitate "recognition" of this specific area (e.g. TF factors) (P. Chambon). Moreover, the gene which is close to this TATA "box" is more often activated by particular regulatory sequences usually located at variable distances upstream of the TATA box. These regulatory sequences can be "recognised" by general protein factors, i.e. which can activate various genes ; such as factors which bind to the CAT sequence (CAT Box). On the other hand, regulatory sequences may have a very specific role in stimulating a particular class of genes. So the "CNNTG" sequence (where N is any base), still called *E-box*, is only recognised by regulatory factors activating genes involved in muscle protein synthesis [1].

---

1. Apart from the existence of these regulatory boxes, the sequences of which are well determined (TATA, CAT, CNNTG, etc.), there are often, far upstream of the transcription promoter, variable and quite long DNA sequences, which interact with particular protein factors, triggering strong activation of downstream genes , sometimes quite distant. We call these DNA activator segments "enhancers": from "to enhance" i.e."to stimulate". Once recognised by the appropriate factor, other DNA segments may inhibit remote transcription: these are "silencers".

It is appropriate at this stage to remember the considerable degree of DNA compaction in eukaryote cells, characterised in particular by the high degree of folding of the double helix. It is therefore probable that transcription factors often intervene by bringing topologically distant areas of the DNA chain closer together in three dimensions. Remember that the transcriptional activation of chromatin also entails remodelling (reversible dissociation) of the nucleosomes, so we can appreciate to what degree gene regulatory mechanisms in eukaryotes are both complex and extremely precise.

I.4.4. CODING AND NON-CODING DNA

So far, our reasoning has assumed that eukearyote DNA is, as in bacteria, comprised almost totally of genetically functional sequences. However, this is not the case, so biologists are confronted with a particularity of nuclear DNA which has been raising many questions since its discovery. In fact, the "genetically functional" fraction of DNA in the chromosomes doubtless represents not more than a few percent of the total DNA (3 to 4%)! (This situation will be touched on again with respect to genomic data). We therefore wonder what other elements form the majority of non-coding chromosomal DNA?

A large proportion of this DNA is certainly represented by introns (see above); the sequences serving as interconnecting elements separating the exons. Yet we must consider these introns to be essential elements of specifically genetic functioning (alternative splicing, extranuclear transport of functional messenger RNA...).

Another fraction is formed, as we have seen, by the many regulatory sequences which act as binding sites for the many protein factors involved in the positive control of gene transcription. But the same comment can also be applied to these regulatory sequences.

Among the other non-coding eukaryote DNA sequences, we find **pseudogenes and repetitive elements.** Pseudogenes are DNA sequences very similar to those of genuine genes, but which are, nevertheless, non-coding, either because they lack start codons, or contain "stop" codons, or are sequences without introns (due to the integration of retroviruses or copies of reverse messenger RNA), the absence of introns having the ability to generate faults in the transcript extranuclear transport mechanism leading to transcriptional copying of these sequences.

I.4.5. Repetitive elements

The repetitive elements of DNA can be extremely varied. They include:

a) Transposons, characterised for the first time by the American biologist, Barbara Mc Clintock, in 1952, in the maize genome, then described in yeast, Drosophila, and finally in all eukaryotes. These transposons are short DNA sequences which, as their name indicates, have the particular characteristic of being able to move from one chromosomal DNA site to another, under the influence of disturbances caused by external factors, such as thermal shock, or genetic reorganisation. Retroviruses also belong to the class of transposable elements. The genomic DNA of endogenous retroviruses is often present in the form of multiple copies which are "mobile". It should be noted, however, that eukaryotes do not contain transposons. Bacterial DNA can also contain some equivalents of transposable elements, represented by episomes (Lederberg, Jacob and Wolleman) or plasmids (mini-chromosomes), which can exist in an integrated state, or a circular, free state inside the cytoplasm and may transfect from one bacterium to another.

b) A whole variety of sequences of variable lengths may be repeated many times. These include some which are interspersed throughout all the chromosomes constituting the genome. They may be short (300 to 500 nucleotides) as, for example, with "Alu" sequences, which appear hundreds of thousands of times and represent up to 3% of nuclear DNA in man! Others may be much longer.

c) Finally, eukaryote DNA includes repetitive sequences, also short, but remaining juxtaposed in "**tandem repeats**". They are often called "**satellites**". These interactive elements are sometimes grouped in specific areas of chromosomes, for example at their ends (**telomeres**), or in constricted zones, (**centromeres**) generally towards the middle. Telomeric sequences, which, through their repetition, often form a long molecular termination at the end of the chromosomal DNA, become shorter with each new division, losing one or more sub-elements. They can nevertheless reform under the influence of specific enzymes, the telomerases, which are less active during the course of ageing. When telomeric sequences become too short, cell division stops.

If, as we see, part of the non-coding DNA can play a more or less direct role in genomic activity (e.g.: introns, regulatory sequences), or replication, we can also imagine that a significant portion of this DNA is involved in the three-dimensional organisation of the genome inside the chromosomes, facilitating the interaction of DNA coils with scaffold proteins.

I.5. GENOMICS – GENERAL DATA – CONSEQUENCES – APPLICATIONS

At the start of the 1980s, molecular biologists began to realise what was available to them in view of the results of research in higher organisms; a highly complex table, the main points of which had been filled in. However, a number of challenges were raised in molecular genetics. The mechanisms which govern the transfer and regulation of genetic information had been almost fully explained. This explains, no doubt, why some of these biologists decided to turn more actively towards development processes, somatic differentiation and embryology as well as the immune response and the neurosciences, using the methods and techniques of molecular biology.

However, apart from the genomes of some bacterial viruses (e.g.: Øx 174) and some eukaryote genes purified by cloning, the chemical sequences of most cell genomes envisaged as a whole, whether of bacterial or higher origins, were still unknown.

The early history of genomics cannot be detailed exhaustively here[1]. To reach their new objective, i.e. to establish the sequences of millions of bases which constitute a genome, scientists resorted to two major techniques, one chiefly chemical, (Maxam and Gilbert) and the other, enzymatic, (Sanger) and in this way they started to "sequence" increasingly complex genomes. The earliest to be elucidated was the bacterium *Hemophilus influenzae* (1985). This success was followed by the characterisation of sequences in various other microbes; yeast, (*Saccharomyces cervisae*), the nematode worm (*Caenorhabditis elegans*), Drosophila, the mouse and finally, at the start of

---

1. The reader can refer to journals or monographs listed in the bibliography.

the millennium, the human genome (at least a provisional version): C. Venter, F. Collins.

I.5.1. Structural and functional genomics

Between a bacterial genome with several million bases, and the human genome, which contains 3.5 billion, or 1000 times more, it was realised that automatic processes, carried out by machines called "sequencers", would turn out to be indispensable. Once the "crude" sequences of a genome are established, it is advisable, for the reasons previously mentioned, to identify those of the "true" genes used in the chemical continuity of DNA. This operation, called "**annotation**", is facilitated by the recognition of elements of promoter sequences, CpG islets, but also of termination and polyadenylation sequences (eukaryote messengers generally contain more or less long chains of poly.A at their 3'OH ends. Introns are located by determining highly specific sequences which form edges at both ends, alternately with the two surrounding exons (Chambon).

But the main objectives of genomics are not limited to the complete sequencing of genomes or basic delimitation of coding and non-coding elements (annotation) of DNA chains studied in this way and counting them, which is the so-called structural genomics approach. They also involve physiologically understanding the roles that these different genes play in cell organisation (functional genomics).

One of the ways of grasping this problem lies, initially, in characterising molecular gene products, i.e. in the study of coding and non-coding RNA as well as, and particularly, identifying their proteins and interactions. Another approach, a kind of characterisation by default, consists of looking for the role of each gene by studying the consequences of its elimination through homologous recombination (so-called "knock-out" technique) (M. Capecchi). We can however surmise that the resulting effect of the experimental obliteration of a gene, will lead both to "specific" effects, due for example to the absence of a particular regulatory factor coded by the gene, and "secondary" effects, at times difficult to distinguish from the previous ones. The knock-out technique is especially useful for specifying the overall physiological consequences of eliminating a particular genetic factor, particularly in the knowledge of its products of expression.

• *The "surprises" of genomics – The number of genes*
      The sequencing of genomes belonging to various species, even various kingdoms, revealed some great surprises once the genes themselves were

counted. It seemed that from one kingdom or species to another, the total number of genes varied only slightly. With 25,000 genes, *Homo sapiens* differs by barely a factor of 5 from the bacterium *E. coli*, very little from the worm *Caenorhabditis*, and even less from the mouse or the plant *Arabidopsis thaliana*! If we now include the genomes of higher plants, we observe that some of them contain practically the same number of genes as man, or even more! (This is the case with rice). The absolute numerical requirement seems, at first sight, to have no strict correlation with the position of the various species on the evolutionary ladder.

Table 1. *Comparative genomic data from bacteria to man*
The DNA in a mammal cell (mouse, human) is 800 times longer that of a cell in the bacterium *E. coli*. The informative content (total number of base pairs) is in approximately the same proportions. However, the number of genes (coding for proteins or RNA) varies in much smaller proportions (about 5 times). This is linked to the fact that, unlike what is observed in *E. coli*, most mammal DNA is "non-coding" (see text).

Organisms	Base pairs (millions)	Total length of DNA	Number of genes	% coding DNA
E. coli	4.6	1.6 mm	4,288	> 90 %
Yeast	12	4.1 mm	6,241	> 70 %
Nematode	97	3.3 cm	19,000	> 27 %
Drosophila	160	5.4 cm	13,600	> 12 %
Mouse	3500	1.2 m	25,000	> 3 %
Human	3500	1.2 m	25,000	> 3 %

This apparent paradox is even more marked when the chemical nature of coding genomic sequences (genes) is examined. Indeed, we note that these sequences are frequently preserved, if not in their entirety, at least in large proportions. This is certainly striking when we compare chimpanzees and humans. The absolute number of genes is identical and the differences observed in their sequences are minimal (1.2%). There are even strong similarites between mouse and human genetic sequences. More striking still is the fact that many yeast genes (with a total genome of only about 6,000 genes) have been retained throughout evolution, across the range of eukaryote genomes, including the human genome. This situation does not imply however that "preserved" genes always fulfil the same function from one species to another, i.e. that the corresponding proteins belong to the same metabolic network (see the hypothesis of so-called "genetic improvisation" put forward by F. Jacob).

How can it be that the number of genes in higher organisms differs so little from one species to another? Several explanations have been put forward. We may think, for example, that a small number of development genes make the difference. These genes might be responsible for the exceptional development of the "telencephalon" in humans (the cognitive abilities of this species, depending much more on learning and education which, biologically, would require preferential stabilization of certain synaptic networks (J. P. Changeux and A. Danchin)) than on the wealth of genes active in the brain. It has also been suggested that evolutionary complexity was determined more by the number of "regulatory" genes or the wealth of regulatory networks as you move up the evolutionary scale. It is clear that a slight difference in the number of elements of a series can generate considerable differences in the number of combinations between these elements. Finally, and especially, it should not be forgotten that the biochemical complexity of a cell or a whole organism is based first and foremost on the proteins which make up its heritage (the notion of a proteome). Alternative splicing mechanisms may lead to a considerable increase in the number of these proteins and most of the properties of cells or organisms depend on their interactions, whether they are structural or metabolic. Whatever the grounds for this interpretation, it is nevertheless true that the situation stemming from the comparative numerology of genes, involving species often very far apart on the evolutionary scale, has led some biologists (K. Scherrer *et al.*) to question the true meaning and definition of the word "gene". To be more precise, it seems that, in higher organisms, the concept of the gene cannot be simply reduced to the molecular definition of its DNA (even by including regulatory sequences). The eukaryote gene is difficult to define outside its functional context since its final products are proteins, which result from splicing, with roles that depend on the interactions between them and other proteins, as well as post-translational modifications (acetylations, phosphorylations, glycocylations, etc.). These considerations will be looked at again when we discuss "systems biology".

I.5.2. GENETIC POLYMORPHISM – SNP

The necessity of enlarging our conception of what a gene is also imposes in the light of studies on the **polymorphism** of genomes. In fact, the establishment of genome sequences of numerous species has revealed a very important new characteristic: the DNA from individuals of the same species presents in its sequences differences or "polymorphisms" which are unique to each individual from one to another and independent from the rare changes due to rather infrequent mutations. These polymorphisms are the source of a true genetic individuality. Thus, the probability that the DNA sequences of two individuals from the same species are absolutely the same, is practically nil. This characteristic is turned to good account in the famous "DNA test" (which has replaced the use of fingerprints) in police investigations or in paternity tests. This individual polymorphism can be manifested under several aspects:

– A particular distribution of the satellite sequences;
– Limited modifications in the genomic DNA sequence characterised by the variation of one unique base pair at the interior of a gene sequence. This type of modification is described in English under the name of "single nucleotide polymorphism", abbreviated to SNP or even "snip", in the jargon dedicated to the people who study genotypes. We consider that on average there is one SNP for every 500 to 1000 base pairs. Thus in the human species, each individual differs from another by several million polymorph traits and, according to the famous aphorism of the geneticist Langaney: "We are thus all related and all different".

The existence of this polymorphism has consequences both fundamental and practical. At the fundamental level, although the presence of a specific SNP at the heart of a gene is not sufficient in the majority of cases to alter significantly its product (conversely to that of a limited mutation capable of changing the properties of the corresponding protein), nevertheless it can bring about changes in the degree of expression of this gene, even in the magnitude of the effects brought about by the mutations themselves. We can then take advantage of the topographic knowledge of the SNPs inside of the genome (**genotyping**), for example to classify groups of individuals according to their SNP. Having established the appropriate statistical correlations, we could for example establish the relationship between this distribution of polymorph traits (notably of SNP) in the genomes of given populations and the probability they have to present this or that phenotype, the degree of susceptibility to cancer or other illnesses. This genotyping of populations is interesting to anthropologists and geneticists. However, it raised fears regarding the emergence of scientific eugenics!

Although at present polymorphism concerns coding genes for the enzymes implicated in the modification of a molecule equipped with pharmacological activity, one will be able to deduce, by correlation research, the degree of therapeutic effectiveness of this molecule according to the polymorphisms observed in this or that group of individuals. It is the principle that is known as **pharmacogenomics**, something which can offer to a certain extent, *à la carte* pharmacotherapy. Whatever it may be, we can readily see by means of these examples that the expression of a gene or the scale of the effects of the mutations can depend to a large extent on the global genomic context which varies significantly from one individual to another.

I.5.3. A BIOLOGY OF MOLECULAR NETWORKS :
TRANSCRIPTOMES – PROTEOMES

If, in its first ambition, the integral sequencing of the genome was aimed at facilitating among biologists a better comprehension of the functioning of a cell by allowing an understanding "en bloc" of the thousands of genes present in the genome of the studied species, a question very quickly arose. This question was about the technical access to the set of RNA messengers (or other forms of RNAs) produced by these genes, and also to the populations of corresponding proteins. Thus, a new step in the study of cell function, a kind of biology of molecular sets, or networks, has slowly appeared.

The complete family of RNAs, produced by transcription of DNA at a given stage of the cell physiology, is called a **transcriptome** and the study of it, **transcriptomics**.

The set of proteins produced from the transcriptome is called a **proteome** and the technical and interpretative method of approach to this proteome is called **proteomics**.

• *Transcriptomes – DNA chips*
The study of transcriptomes has been made possible for the first time in history by the introduction of a new technology described in 1954. There, the authors reported the hybridisation of the entire pool of the "transcripts" (RNAs) derived from the plant *Arabidopsis* (after conversion of the RNAs by reverse transcriptase into c-DNA) with short DNA sequences, corresponding to well-defined portions of known genes from the plant, immobilised on a solid support. Thereafter, the immobilised DNA probes on a support (strip of glass or silicium) were composed either of short DNA chains, obtained by amplification due to the PCR technique (Mullis) or by the oligonucleotides synthesised directly on the solid support by microphoto-lithography (Affymetrix technique), these synthetic oligonucleotides reproducing the characteristic sequences of known genes. In practice, hundreds, even thousands of probes of DNA from known genes or from oligonucleotides reproducing their sequence, are then either deposited or synthesised *in situ* and arranged on the

solid surface in a set order and according to a well-determined topology, so that the RNA messengers which make hybrids with these probes can be easily characterised. In general, before depositing the cell extract containing the global pool of RNA, these RNA transcripts of genes are converted into c-DNA (complementary sequences of DNA obtained *in vitro* by the inverse transcriptase) and marked with a fluorescent agent. One can then detect the fluorescent spots corresponding to the hybrids formed. In fact, automatic procedures allow them to be detected and in this way, transcripts initially present in the extract can be identified (and to some extent, quantified).

This technique based on the utilisation of "DNA chips" allows us to specify what genes are expressed in a cell (transcription) at a given instant (gene profiling technique). Since 1990, there has been a great number of applications. For example, this technique permits one to compare the spectrum of gene expression in a normal and cancerous cell (a useful approach for establishing the prognosis in the evolution of certain cancers). It is also used by people interested in population genetics. It gives some precious information concerning the networks of the genes that are expected according to the state of development of a tissue. It can reveal the modifications of the transcriptome caused by a pharmacological agent or a hormone or even changes in the register of gene expression, following a genetic disease, etc. However, although the technique of DNA chips allows very interesting qualitative studies, it still does not lend itself easily to measuring precisely the rate of gene expression.

• *Proteomes*
Regarding proteomes, great progress has also been achieved in their comprehensive study. However, the difficulties here are linked to the great diversity of the proteins potentially expressible in a cell, notably in a eukaryotic cell. In fact, by the process of alternative splicing, it has been calculated that on average each eukaryotic gene codes for 5 to 10 proteins of sequences, related but distinct. To this should be added the multitude of variants resulting from post-translational modifications: acetylations, hydroxylations, glycosylations, phosphorylations, etc. It is therefore probable that the potential diversity thus realised can approach several hundreds of thousands of entities or even more! The techniques implemented to analyse proteomes are varied. As has been achieved in the study of transcriptomes, one can make use of "biochips". For example, one can have recourse to the use of solid supports immobilising the diverse protein "ligands" (chemically synthesised agents, chemical analogues of substrates, proteins, nucleic acids, etc.). The protein extract to be analysed is layered on the micro-network, where each immobilised ligand retains the protein to which it displays the highest affinity. Certain techniques allow the analysis of the reactivity of the proteins thus retained. For some time, the

capture of elements of the proteome has been achieved by using specific antibodies with the appropriate affinity as immobilised ligands.

However, one of the most used techniques is based on two-dimensional agarose gel electrophoresis, a standard technical method implemented for decades, and which has known numerous improvements. The proteins subjected to electrophoretic separation are located by autoradiography after being marked by a tracer. After elution (separation) from the gel, and submission to proteolytic cleavage, their peptide sequence can be determined by mass spectrometry.

A considerable number of protein sequences has thus been established and their description recorded in large databases (e.g. Swissprot). However, although knowing the sequence of numerous proteins from different species might give rise to the study of interesting comparative studies, it only uncovers part of the information about them. True artisans of the metabolism, of the physiology of a cell or of immune defences, but also chief components of the sub-cellular architecture and of numerous organels, the proteins owe their great importance to their aptitude for achieving a "stereospecific recognition" of very many ligands at the level of their active sites, and to the multiple interactions which they are susceptible to establishing between them.

Knowing their three-dimensional structure and the network of interaction in which they participate also constitutes one of the major goals of proteomics.

Since the pioneering work of M. Perutz and J. Kendrew on the three-dimensional structure of haemoglobin, multiple proteins have seen the establishment of their spatial configuration and the clarification of their mono- or multimeric nature (one or several sub-units). This is indeed long-term work involving crystallisation techniques and the precise determination of spatial distribution of different component atoms, making use of X-ray diffraction or spectrographic techniques.

Taking into account the enormous diversity of proteins and the technical difficulties associated with the establishment of their structure in space, methods capable of predicting their complex 3-D folding from their primary sequences are indeed of major importance. Great progress has been achieved in this regard and the existence of certain "peptide motifs" is often revelatory, not only of the tertiary architecture of the protein which contains them but also of its physiological function. For example, typical three-dimensional configurations are found among the proteins, acting as positive transcription factors in the eukaryotes. Some of the "structures" that can be encountered are called "zinc fingers", where the zinc atom allows folding of the peptide chain in the form of "fingers", which thus come into contact with a regulatory DNA sequence, lying upstream of a promoter (A. Klug). Others are designated as leucine-rich zippers. Other conformations of regulatory proteins respond to the arrangement called b-HLH, where the protein comprises a sequence rich in basic amino acids (b), preceding an arrangement in which two helicoidal constructions (a-helices, H) are separated by a loop (L: "loop"), the basic part facilitating the electro-static attachment of the factor to DNA, while the whole HLH moiety causes its interaction with other factors also bound to DNA in the vicinity of the promoter.

• *Protein interaction*

However, as we have previously pointed out, another challenge which is in reality from the **post-genomic era** is to understand how the expressed proteins **interact** with each other to allow the expression of a defined physiological function within that particular "micro-environment" which a cell is. In other words, from now on, contemporary biology takes on a **level of complexity** which molecular biology had not even attempted to attain. Thousands of proteins are expressed in the cell or the tissue at specific stages of their development. Now, although each protein taken in isolation can be assigned a particular activity (enzyme, receptor, signalisation factor, chaperonin[2], etc.), there is every reason to believe that in the extremely compact environment which a cell represents, the proteins act, *in vivo*, according to physico-chemical constants (affinities, speed, etc.) distinct from those observed *in vitro*; above all, each protein participates in a complex network of "protein-protein" interactions. Specifying the network of interactions which are underlying specific metabolic function should open the way for an integrative biology of living things (often referred to as metabolomics).

To the attempt of characterising complex networks of protein interactions are also attached those aiming at specifying their location in the intracellular micro-space (thanks to confocal microscopy, for example) and the objective of explaining the dynamics which presides at their delivery (the intracellular flow of proteins). Varied techniques already used for several years have allowed protein-protein interactions to be brought to the fore. For instance, one can have recourse to a trap system by a specific antibody: as an example, a protein "A", the partners of which one tries to identify within

---

2. Chaperonine: protein capable of associating specifically with another protein and of guiding the latter towards a defined cellular micro-compartment so that it allows it to exercise its activity.

a defined network of interactions, will be covalently fused with protein "B" by genetic engineering, protein B serving here as a kind of "label" or tag. The soluble cellular extract (containing protein A) is then passed through a column holding an antibody directed against the tag. Only the proteins present in the extract which possess a strong affinity for protein A will thus be retained. The link between protein A and its tag is then split using appropriate enzymes. As a result, the complex formed between protein A and the many other interactive proteins will be "desorbed". By electrophoresis under denaturing conditions, the various proteins that are naturally complexing proteins will be separated and characterised.

Another very widespread technique is that of the "double hybrid". We have seen that an activation factor of type b-H-L-H encompasses several sub-domains: one, for example, serving to fix the factor to DNA while the others stimulate the transcription by acting in the immediate vicinity of the promoter. For simplification, let us call "α" the first domain and "β" the second . The factor consisting of the domains $\alpha + \beta$ is functional (it can activate the trans-criptases), whereas each element taken separately is not. At present, let us suppose one suspects two proteins, P_1 and P_2, of being capable of interacting with one another. By genetic engineering, we can construct a fusion protein, α-P_1, and a distinct fusion protein, β- P_2. Both artificial "constructs" can then be transferred into a cell. Actually, if P_1 and P_2 do form a natural complex in the cell, the domains α and β will be artificially reassembled, which, in numerous cases, will reconstruct the stimulatory effect of the factor α-β at the transcrip-tional level. One will thus be able to deduce from this that, in the cell, proteins P_1 and P_2 naturally interact with each other. Step by step, this double-hybrid technique has made it possible to draw up a map of the numerous protein interactions taking place in a unicellular organism, such as *B-subtilis* or the yeast *Saccharomyces cerevisiae*[3].

• *From genome to phenome*

At this stage of our discussion on genomics and post-genomics, what general conclusions can we draw? During the phase which preceded the development of genomics, molecular biology, in its ambition of explaining cellular function, was mainly interested in studying typical (illustrative)

---

3. This technology, interesting and innovatory though it may be, will only take on its full value in so far as it is possible to show that, among these multiple interactions, some are directly involved in a metabolic or signalling pathway. In this regard, we can note that if this novel approach to mapping the protein interactions can unravel integrated cellular function, bioche-mistry has already succeeded in bringing to light true signalling cascades (of protein to protein), involving protein kirases and protein phosphatase (E. Fischer).

situations often taken out of the context of the intra- and extra-cellular global environment.

For example, the analysis of genetic regulation has, for a long time, benefited from the studies of the lactose system of *E. coli* (induction and repression of the lactose operon) or of the system represented by the "induction" of prophage λ. These approaches have been rewarding.

However, with the development of genomics and the fine, detailed study of transcriptomes, proteomes and (soon that of metabolomes), biologists have begun to tackle the functioning of the cell factory by large groups (or networks). As a result, little by little, it was realised that these "sets" did not constitute isolated compartments. In particular, the number of proteins belonging to the proteome (and even as we shall see later, the number of non-coding RNAs belonging to the transcriptomes) either interact with the DNA sequences or with elements of the transcriptome. In other words, each set interacts with the other. The highly integrated nature of the "compartments", or cellular sub-sets, imposes itself in an even clearer way from the instant when we begin to address such phenomena as physiology and somatic differentiation, whose levels of complexity are greater than those of an isolated cell. For example, several metabolomes can intervene in what some already call **physiomes**, even **phenomes** (i.e. all phenotypical manifestations on the scale of the whole organism). The functioning of the cell by way of more or less closed "loops" and the awareness of its various levels of functional integration within an organism are gradually leading to the formation of two main approaches.

I.5.4. WHAT IS A GENE? SYSTEMS BIOLOGY

The first current of thought adopted by some biologists is related to the hierarchy attributing to the genes a unique role in the determination of the cellular functions. DNA is a text which does not make sense if it is not decoded by a reader (in the instance of a cell). In other words, the code cannot be "read" concretely (that is to say, cannot be transposable into integrated cellular functions) without the joint actions of the cell and of its environment.

The second has to deal with methodology: it rests on the idea that without a new Biology, called "Systems Biology", based upon very powerful tools for conducting analyses and forecasts, integrating Information Technology, nanotechnologies, and in some cases high resolution physics and mathematics, it would be vain to hope to embrace the cellular complexity, the physiological functioning of an organ or that of an ecosystem, and this, despite the considerable amount of data from biochemistry, from molecular biology and the diverse approaches due to genomics. As Pierre Vignais writes (Experimental Science and Knowledge of the Living, p. 320, 2006): "from a mono-parametric approach, which at the beginning was often essentially and necessarily reductionist, the experimental method applied to the living has come to an approach which is "globalised" or synthetic, multiparametric and whose ambition is the comprehension of molecular interactions in defined biological sequences. In making use of obtained data, the hope is to simulate, by a mathematical treatment, the global functioning of a cell, an organ or an organism". As one can understand, the new systems biology's ambition is not limited to a <u>description</u> of the integrated functioning of the cell. Its originality is equally to formulate some predictions relative to changes in the behaviour of the system depending on the manipulation of various parameters (concentration of substrates, presence of inhibitors, etc.).

This objective is certainly difficult to attain. Consider the case of enzymes: it is not sufficient to know their specific substrates and the Michaelis constants to predict how they function in the cell. Indeed, it is clear that their degree of compaction and of common ownership will be able to change considerably their effectiveness. It remains that by considering the complexity of biological systems from the "modular" angle, one can treat the whole of the interactions which intervene in this module by mathematics and, from these elaborate models, draw some interesting predictions.

Another difficulty, by no means the least, is in the existence of **stochastic** phenomena. For example, this is the case of precursor cells in the process of differentiation, which have two choices; either to differentiate themselves or to remain in their initial state (B. Laforge *et al.*).

In 2004, the International Union of Physiology, which is one of some thirty international scientific unions of the ICSU (the International Council for Science) launched an ambitious programme of systems called "Physiome", and there are already hundreds of publications in favour of this new approach for predictive modelling of living organisms.

These new ways of approach perhaps offer the prospect of seeing little by little the emergence of a new biology better able to integrate, in its description of the phenomena, what precisely constitutes the complexity that is proper to living organisms.

I.6. A NEW INSPIRATION IN MOLECULAR BIOLOGY – THE WORLD OF RNAS AND THE PHENOMENA OF INTERFERENCE – THE RETURN OF EPIGENETICS

I.6.1. THE WORLD OF RNAS

In science, there is often an illusion of having exhausted certain areas of knowledge. We think that "everything has been said" regarding this or the other system of study, as it seems that the amount of data collected on the subject is logical and exhaustive. However, nature often reserves surprises. These often call into question a whole body of doctrines. Biology has been particularly productive in these kinds of "rebounds". We must recall the discovery of the inverse transcriptase, that of mosaic genes and of splicing mechanisms, etc., or more recently that of prions.

About a decade ago, while the overall scheme describing the genetic regulation at the heart of the eukaryotes had clearly established the existence of positive control factors (that is to say, of specific proteins acting at the level of regulatory sequences, located close to the transcription promoters present in the DNA), diverse observations, firstly isolated, then more and more numerous, finally brought to light a totally unexpected phenomenon linked to the action of a new category of RNA molecules. Then, what appeared as an exceptional situation turned out to represent a rather general mechanism of regulation. It was found that in the cell, next to the RNAs which have been known for a long time (ribosomal RNA, transfer RNA and messenger RNA), there exists a new family of small RNAs whose principal role is to block the translation into proteins of messenger RNAs by pairing with some complementary sequences. Sometimes they even provoke their destruction. These small RNAs, smaller in size than transfer RNAs, behave like true repressors of genetic translation! This mechanism has been named: **interference** and the RNAs which are responsible for it: **small interfering RNAs (Si-RNAs)**.

I.6.2. Si-RNA AND MICRO-RNA

At present, one can describe two categories of RNAs capable of causing interference with the functioning or metabolic survival of messenger RNAs:

- Small interfering RNAs (abbreviated to "siRNA"), so called because they can interfere with the translation of certain messenger RNAs, and most often cause their destruction.
- Micro-RNAs (miRNA) whose physiological importance is gaining increasing support.

The siRNA come from long RNA "duplexes" or ds-RNA (double-stranded RNA) resulting from the activity of cellular enzymes called RNA-dependent RNA polymerases. These long ds-RNAs can be formed in the cell from "transposons", from viruses, from DNA present in the heterochromatin, etc. The ds-RNA thus formed is converted into si-RNA (short sequences of 21-23 nucleotides). Then, one of the two strands of the ds-RNA associates with a messenger RNA carrying a complementary sequence. In most cases, the corresponding base pairing association is cleaved in its middle by a nucleo-lytic enzyme. These multi-step operations are mediated by complex systems of transport and recognition which are described below. These siRNAs, unlike the miRNAs, are not conserved at the evolutionary level. They are formed occasionally from dsRNA in particular circumstances. The Si-RNA-mediated interference phenomenon has been experimentally identified for the first time by A. Fire and C. Mello by injecting some double-stranded RNA into the nematode worm (*C. elegans*).

As already mentioned, micro-RNAs, unlike the siRNAs, have been conserved during evolution. These are the products of specific regulatory genes, particularly abundant in eukaryotic organisms. These genes are often arranged in pairs in the DNA. The existence of these micro-RNAs was reported for the first time in 1993 by Lee *et al.* (1993). These authors observed that one of the regulatory genes, Lin-4, intervening in the control of the early stages of the development in the nematode worm *C. elegans*, did not produce a protein as one would have expected, but a short sequence of RNA! This antisense RNA was shown to act by inhibiting, in a post-transcriptional way, the development gene lin-14. This observation, fairly surprising at the time, was reinforced and clarified several years later by A. Fire and C. Mello (Nobel Prizewinners 2007). In a series of experiments using the injection of double-stranded RNA into the nematode, they could demonstrate that one of the two RNA strands acts by blocking the RNA messenger transcribed from another regulatory gene, unc-2. In retrospect, this interference mechanism also provided the explanation of results obtained at the end of the 80s by R. A. Jorgensen who, paradoxically, noticed the bleaching of petunias which had been treated by "RNA copies" of genes involved in the synthesis of the purple pigment!

Later, Reinhart *et al.* (2000) highlighted another regulatory gene, lin-7, acting at a different developmental stage of the nematode by producing, like lin-4, a 21-nucleotide-long antisense RNA. Similar observations indicating that short antisense RNAs are involved in the larval development of *C. elegans* have also been be reported by various authors between 2000 and 2004 (for an overview see D. P. Bartel and C. Zheng Chen (2004)).

Little by little, it turned out that regulation mechanisms involving small antisense RNAs are an extremely widespread phenomenon. Small antisense RNAs are capable of interfering with the RNA messengers issued from key genes of the development. This mode of regulation has been observed in situations as diverse as the control of the cellular proliferation, cell death and fat metabolism (in Drosophila), the morphogenesis of the nervous system in the nematode or the control of the formation of the leaves and flowers in the *Arabidopsis* plant. Also, as we have emphasised, while the regulation system resting on the action of the miRNAs is conserved in mammals, one can equally observe its occurrence among invertebrates. Very many target molecules of micro-RNA have been identified in the last few years, among both plants and animals. In 2003, B. P. Lewis, D. Bartell *et al.* had already identified more than 400 such molecular targets among the vertebrates.

• *Formation – Transport and matching of micro-RNA*
(Drosha, Exportin, Dicer and Risc)
    The mode of transport of micro-RNA precursor molecules, from the nucleus to the cytoplasm, their conversion into active micro-RNA and their mode of fixation and assembly at the level of target messenger RNA, depend on a very complex cellular machinery which is beginning to be made clear (cf. E. P. Murchison and G. J. Hannon, 2004). The principal protagonists of these reactions are entitled: <u>Drosha</u>, <u>Exportin</u>-5, <u>Dicer</u> and <u>Risc</u>.

    <u>Drosha</u> is a type III ribonuclease (i.e., one of multiple ribonucleases which catalyse the split of double-stranded RNA). The enzyme is present in the nucleus of mammalian cells (men, mice etc.). The enzyme's role is to cleave

the product of primary transcription of the micro-RNA coding genes or "pri-miRNA". This pri-miRNA presents itself as a folded-hairpin-like structure within the nucleus; its cleavage engenders a secondary precursor of micro-RNA, or "pre-miRNA", another hairpin structure of around 70 nucleotides.

Exportin-5 intervenes in the interior of the nuclear compartment to transfer the pre-mrRNA thus formed into the cytoplasm.

Dicer, another RNAse III, converts the pre-miRNA into a 22-nucleotide micro-RNA inside the cytoplasm. Dicer is a modular enzyme, comprising two "RNA helicase" domains, from a module named PAZ and from a motif involved in the recognition of double stranded RNA. The PAZ module is essential for the recognition of the pre-miRNA (at the level of its 3'end).

"RISC" (an acronym for "RNA-induced silencing complex") supports the mature micro-RNA resulting from the enzyme conversion by DICER and carries it to its target-messenger RNA, by facilitating the recognition of the complementary sequences present in one of the two strands of the micro-RNA and in this messenger. (This process of "apposition" of the mi-RNA onto the messenger target thus involves a cleavage of the single strand loop of the mi-RNA and the separation of the two strands which constitute the stalk.) There are several types of RISC complexes varying according to the species. Generally, there are ribonucleoproteins containing protein constituents, called "Argonauts" (AGO) and numerous stimulatory factors whose nature varies according to the species. "Argonaut" proteins are supposed to allow the incorporation of the mature mi-RNA into the RISC system[1].

According to the case, once hybridised to the complementary mi-RNA strand, the "target" messenger is either cleaved or its translation into proteins is inhibited. The intracellular fates of si-RNAs and mi-RNAs are identical. They involve the same intermediary steps.

----

1. Among the plants, aside from 21-nucleotide-long si-RNAs (involved in the resistance to pathogens or in the inhibition of transgenes...) which constitute a sort of immunity, one can encounter another category of 24-nt-long si-RNAs. The latter are involved in a complex called RITS (RNA-induced transcriptional silencing). This RITS complex, like RISC, carries an Argonaut protein. However, this is combined with several nuclear proteins. The 24-nt si-RNAs act in "CIS" and are essentially involved in the control of movement of the transposable elements and in the maintenance of the heterochromatin structures (H. Vaucheret, common session, Academy of Science/Academy of Agriculture on "Riboregulators", new dynamic actors of development, 5 December 2007).

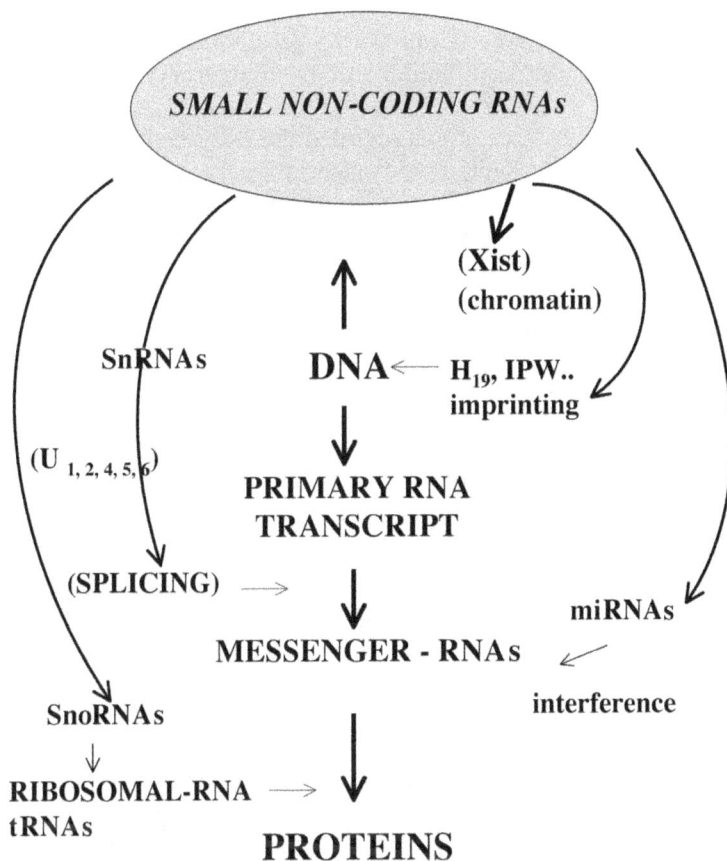

Fig. 5 *The "new world of RNA" – role of small "non-messenger" RNAs*

Diagram showing the wide diversity of roles played by small non-coding RNAs in cell physiology:

rRNA: Ribosomal RNA; tRNA: transfer RNA (amino acid adapter).

Xist: RNA with about 50 nt (nucleotides) involved in X chromosome inactivation (see text: "epigenesis").

H-19: involved in the so-called "parental genetic impregnation" process.

MiRNA: micro-RNA, (21 to 23 nt): provides negative regulation (like siRNA: silencing RNA) on messenger RNA transcription by forming a molecular hybrid RNA: RNA.

SnRNA: Small nuclear RNAs (types U, 2, 4, 5, 6) involved in splicing. These are the components of "spliceosomes".

SnoRNA: Small nucleolar RNAs.

RNAs of between 60 and 300 nt, involved in ribose methylation of rRNA or in the pseudo-uridylation of tRNAs.

• *Applications*
    The discovery of interference RNA (si-RNA and mi-RNA) has had important consequences, both fundamental and in terms of application:

    – in the first instance, it has revealed the existence, in the eukaryotes, of a new and very remarkable post-transcriptional regulation system which would have appeared late in the course of evolution, since it has not been observed so far among the prokaryotic organisms.

    RNA-dependent interference phenomena are very widespread among plants, notably in processes leading to the inhibition of viral or viroid activities (a kind of plant-specific immunity). R. Plasterk has shown that they also play a major role in animal eukaryotes by inhibiting the invasive role of transposons. From a more general standpoint, biologists' attention has recently been focused on the roles RNAs must have played in the past – and are still playing today in the economy of the cell and in developmental processes. For example, the properties that some RNA molecules can harbour while mimicking the activity of enzymes (ribozymes) have already been reported. Recently, it has been shown that, in the genetic translation process, the principal stages of elongation of the polypeptide chain which take place at the level of the ribosomes are not catalysed by the numerous ribosomal proteins of this complex organite, as had been thought, but by the ribosomal RNA which functions as a true ribozyme (A. Yonath).

    Since the 50s, the essential role of the transfer RNA in the recognition of the codons of the messenger RNA and in the "positioning" of the amino acids has been understood. RNAs are equally involved in a large number of cellular activities. The **spliceosomes,** organels which, as we recalled, promote the excision of the intron transcripts in the course of the splicing process, require in their activity a special class of small RNAs, or Sm-RNAs (59 of which have been described). Other small-sized RNAs (like the Xist-RNA) have equally been described as playing an essential role in the inactivation of the X chromosome or at the level of the heterochromatin. From these various observations, biologists have begun to forge the idea, getting more and more credence, that the world of RNA, whose role had been partly under-estimated until now (probably because it had been eclipsed by that of DNA) will most likely still reserve some more surprises! Last (but by no means least), chemically synthesised interfering RNAs are the object of a growing use. On the one hand, in fundamental research, the study of the physiological role of numerous genes of interest (functional genomics) is very often achieved by artificially blocking their transcription by means of synthetic double-stranded RNA. On the other hand, modern pharmacopoeia is becoming gradually enriched by numerous synthetic oligo-ribonucleotides susceptible to intervening as therapeutic agents. The treatment of viral infections (such as AIDS, hepatitis B, etc.) could possibly benefit from these new technologies.

I.6.3. THE RETURN IN STRENGTH OF EPIGENETICS – WHEN HETERODOXY BECOMES A SYMBOL OF OPENNESS[2]

If there is one "fashionable" concept today in biology, it is probably that of epigenetics because it meets with a certain aspiration for a more synthetic and integrated knowledge of the development and the physiological functioning in higher organisms.

The term itself has many meanings and keeps acquiring new ones as research makes progress on early development (for example, root cells and nuclear reprogramming), on the adaptability of living organisms to their environment, or even on certain of their dysfunctions (e.g. cancers). This might explain the renewed interest of researchers in life sciences for what is on the way to become a true discipline, although one still with many difficulties regarding its precise definition. The word "epigenetics" was introduced and its concept forged 6 decades ago (1942) by Conrad Waddington. Paradoxically, while epigenetics initially responded to a concern for defining the intervention of a complex set of genes in the embryonic development, little by little it reoriented towards liberating developmental biology from pure molecular "Jacob-Monod" style genetics.

In fact, the classical schema of a regulation at the level of the transcription of the DNA, to a large extent founded by the study of the adaptation of bacteria to their environment although partially transposable to eukaryotic organisms, turned out to be insufficiently adapted to the explanation of processes as complex as embryogenesis and somatic differentiation. From then on, epigenetics would not only be concerned with the functional contribution of the vast network of genes in the development and in the physiology of major functions, but it would try to accord a much bigger place to the effects of the environment in hereditary transmission (without, however, going as far as to revitalise neolamarkism).

If this new meaning and this trend continue to prevail – on the background against a of a more or less explicit rejection of classical genetic determinism seen as too strict – nevertheless biology focuses today, above all

---

2. See Biofutur n° 243, p. 18-31 (2004).

on the phenomenological aspects of the problem. One of the recent definitions resulting from this new emphasis as follows:

"(Epigenetics is) the study of the changes in the expression of the gene which are heritable, at the time of the mitosis and/or the meiosis, and which do not result from modifications of the DNA sequence" (V. E. A. Russo, R. A. Martienssen and A. D. Riggs, in "Epigenetic mechanisms of gene regulation", Cold Spring Harbor Laboratory, press, p. 1, (1996)). More simply, one can say that epigenetics concerns the modes of transmission of the specific cell characteristics that are not genetically coded.

Epigenetic manifestations which, up to now, have received the most attention, as they have been the object of the most detailed study, are related to the methylation of certain strategic sites of the DNA, but also to biochemical changes (more or less reversible and under the influence of the environment) at the level of histones, some main constituents of chromatin, in circumstances such as: the activation or functional "locking" of large parts of the chromosome, the inactivation of the X chromosome in mammals, or in the phenomenon called "parental imprinting".

• *DNA Methylation*

Among the most current epigenetic modifications, capable of leading to heritable changes in the functioning of certain genes, figure the DNA methylations.

These methylations are susceptible to taking place at the level of the bases A or C, and the methyl group thus attached is localised in the large groove of the double helix which influences its binding capacity for proteins.

The donor of the methyl groups is a universal molecule which has been universally characterised more than half a century ago (G. Cantoni, cf. Schubert *et al.*), S-adenosyl-methionine. Regarding methylation enzymes, the methyltransferases, they belong to two categories: some are capable of de novo methylating the bases A or C of DNA; other categories only function if one of the two strands of the double helix is methylated beforehand.

The process of DNA methylation must have appeared early in the course of evolution. It is already observed in prokaryotic organisms such as bacteria. It intervenes in the phenomenon called "restriction". In fact, the DNA of a defined bacterial species is methylated in strategic sites, which gives it a particular "immunity" vis-à-vis nucleolytic enzymes. If a DNA virus (bacteriophage) infests the host bacterium, its non-methylated DNA will be destroyed.

Curiously, DNA methylation has not been observed, or only in an exceptional way, in certain eukaryotic organisms such as yeast, nematodes and drosophila, while it is common among mammals. The latter contain two methyltransferases able to intervene in different circumstances. DNMT-3a and 3b can modify the DNA *de novo*, while a third methylase, DNMT T-1, can only act if one of the two strands of DNA has previously been methylated.

Among mammals, methylation only affects a single chemical motif at the heart of the DNA chain. This is the dinucleotide CG (also called CpG island). This dinucleotide is principally found at the level of DNA sequences which act as start sites (or promoters) of transcription. When the cytosine C of the dinucleotide is thus methylated, the promoter site becomes inaccessible to the transcriptase. Indeed, specific transcriptional repressors of the MBD family are fixed on the methylated cytosine and "recruit" another protein, the enzyme histone deacetylase (HDAC). When the histones are not or are no longer acetylated, the interaction with the DNA of the nucleosome (of which they are constituent elements) is considerably reinforced. By contrast, the methylation of the CpG islands of the promoter can block, in a permanent and heritable way, the functioning of the genes which are under the dependence of the methylated promoters.

Such permanent genetic modifications, following DNA methylations, are equally seen in diverse situations such as: the transcriptional repression of transposons, the process of parental imprinting or the inactivation of the X chromosome.

• *Transcriptional repression of transposons*
    As already recalled, transposons are "mobile elements" whose sequences are often repeated along the DNA chain. They can represent up to 45% of certain genomes. They propagate often in the course of cellular divisions (this involves making an intermediary RNA, resulting from their transcription), and this propagation can affect the active regions of the chromosome and provoke a harmful interference in the functioning of numerous neighbouring genes. Therefore, the cell has developed a mechanism allowing it to avoid or restrain this phenomenon. To this end, it makes use of interfering RNAs, to counteract the effects of the transcription products derived from the transposons, or it initiates massive processes of methylation of the transposons themselves, which block their transcription, and hence their propagation.

• *Parental imprinting (differential expression of certain genes of paternal or maternal origin)*
    The phenomenon of parental imprinting (or impregnation) takes place when certain genes, around fifty in man, playing for the most part a role in the development of the foetus, must be quantitatively regulated in a very precise

fashion. This is the case, for example, in the coding gene for the factor IGF-$_2$ (insulin growth factor type II) or H$_{19}$. To achieve this aim, these genes are only expressed at the level of one of the two copies of chromosomes, which in the somatic cells of the foetus are inherited from the parents; sometimes these will be the maternal chromosomes, sometimes the paternal ones. Here again, the permanent inhibition of the genes on one or the other of these chromosomes results from a methylation of their promoters, leading to a transcriptional repression.

• *Inactivation of the X chromosome*
    Female somatic cells have a caryotype XX, and those of males are of type XY.
    There are around 1,200 genes in the X chromosome. If the epigenetic rearrangements did not intervene, the female cells would express the equivalent of twice this number, against only one equivalent among the male's. In order to reequilibrate this "genetic dosage", one of the two X chromosomes of the female somatic cells undergoes a complete functional inactivation in the early phase of the female embryonic development, as was brought to light for the first time by Mary Lion in 1961. The inactivation results in a transcriptional repression at the level of all the genes present in one of the two X chromosomes. The nature of which of the two X chromosomes undergoes inactivation responds to a random process; it can correspond to an X chromosome inherited from either the father or the mother. (From this point of view, it is interesting to realise that there are two cellular populations in the embryo).

    The mechanism involved in the selective inactivation of the X chromosome has been the object of numerous studies. They can be summarised as follows:

    The triggering of the inactivation process occurs at a single locus or inactivation centre (Xic). Among the genes present at this locus, the most significant one (P. Avner, E. Heard , K. Plath *et al.*) codes for a small RNA molecule, called Xist. This is a non-translated 17kb RNA, whose expression increases at the early phase of differentiation. Xist RNA recovers two X chromosomes at the beginning but, very quickly, the gene coding its synthesis is repressed on the chromosome that will remain active. The RNA Xist not only triggers the inactivation of one of the two X chromosomes but also enables the propagation of the inactivated state starting from the Xist locus. By following the process in ES type embryonic cells of the mouse, during *in vitro* differentiation, one can clearly establish the existence of a first stage of inactivation that is reversible (if the gene coding for this RNA is repressed), followed by a second, irreversible and Xist-dependent stage.

During the reversible Xist-dependent phase, one can observe modifications in histones H_3 and H_4 (methylation and hypoacetylation) followed on the 2nd day of differentiation of ES cells, by a definitive transcriptional repression. Towards the 3rd day, the inactive X chromosome is replicated in an asynchronous way compared to its active counterpart. Then, the irreversible inactivation phase begins, independent of Xist; then, on around the 7th day, the presence in the nucleosomes of a variant of histone H_2A, known as macro-H_2-A_1, can be observed. Methylation of the CpG islets, in the vicinity of gene promoters, finally intervenes to ensure the maintenance of the inactive state.

Recent data (J. Silya and K. Plath) has brought relative precision to the reversible phase of the inactivating process. It indicates that the accumulation of the RNA Xist on the chromosome destined to be inactive involves the recruitment of a complex protein of the "polycomb" group, Eed/Enx-1, which would be involved in the methylation of lysine residues 9 and 27, within histone H_3 (cf. above).

• *Histone modification and the role of histone variants in epigenetic control*
    The inactivation of the X chromosome illustrates, as we have just seen, a process of transmitting epigenetic modifications via histones, in particular during the phase corresponding to the definitive "locking" of the chromosome. In fact, this locking is "perpetuated" during successive divisions of the somatic cells which follow the differentiation of the embryonic cells; the return to the active state is only produced after the meiosis preceding the chromatic reduction.

More generally, the histones' makeup of the nucleosomes can undergo diverse modifications and, most importantly, can be replaced, during development, by histone "variants", whose assembly gives to these nucleosomes the capability of "flexible" interaction with DNA according to the stages of this development. Histone variants (of which we have just encountered an example with histone macro-H_2A_1 – see above – during the irreversible inactivation of the X chromosome) thus participate in the structural and functional definition of the genomic domains, and are therefore vectors of genetic information in the same way as their post-translational modifications.

However, in order to fully understand the role which the variant forms of histones can play at the epigenetic level, we must return to several generalities regarding the dynamics of the assembly of the nucleosomes. One remembers that these are formed by an octamer of "major" histones with the formula $(H_3,H4,H2A,H2B)_2$, the octamer representing the core of the nucleosomic particle.

Chromatin, a compacted set of DNA and nucleosomes, consists of a mixture of parental histones and of histones that are newly synthesised during the S phase of the replication cycle, the latter having to possess the same epigenetic marks as the parental histones in order to allow the epigenetic information to be maintained through the successive divisions. The assembling of the nucleosomes on contact with the DNA involves in the first place the fixation of the dimer H_3-H_4, followed by that of histones H_2A and H_2B. The histones are transported from formation site, by dedicated proteins specific to each of them, and which are responsible for their cellular fate, in the sense that they become attached to them and carry them up to the DNA. (These proteins, which in this way ensure the intracellular traffic of other key proteins, are called "chaperon" proteins or "chaperonins".)

Besides these four types of major histones described above, numerous isoforms, called "replacement histones" or histone variants have been described (Franklin and Zedler). Only histone H_4 does not possess any variant. H_2A, H_2B and H_3 each exist under two forms called 1 and 2 (for example, H_2A_1, H_2A_2, etc.) as major histones. In man, four histones of replacement of H_2A_1 and H_2A_2 have been identified; namely, H_2AX, H_2AZ, MacroH$_2$A and H_2ABbd. Replacement histones corresponding to forms H_2B_1 and 2 have also been described; namely, TSH$_2$B and H_2ABbd, and finally, 2 histones of replacement of H_3 and 2; namely, $H_{3.3}$ and CENP-A.

The replacement of the major histones present in the nucleosomic core by certain of their variants gives new properties to the nucleosomes and hence to chromatin. Thus, although they are minorities, these variants can play a major role by participating in specific functions of the cell. We have seen the example of the form MacroH$_2$A which becomes preponderant in the inactivated X chromosomes while it is absent in the active ones. (On the other hand, H2ABbd is excluded from the inactive form.) For another example, the variant H_2A-X in its phosphorylated form would be associated with chromatin following double-stranded breaks in the DNA (a possible role in repair) while H_2A-Z is concentrated within the nucleosomes in the chromatin of embryos, during the course of their early development. The variant CENP-A figures in a specific way in the chromatin of the "centromer" by replacing H_3 and its presence is associated with the attachment of the kinetochores[3]. $H_{3.3}$, another variant of H_3 replaces this major histone in cells while in their non-proliferative phases. Its presence in the place of H_3 would somewhat "relax" the dimerisation of the couple H_3- H_4 and this would favour the interaction of the transcriptases and the DNA.

As previously reported, the different isoforms of histones, as well as their major representatives, are put in place by chaperon proteins. Thus, protein CAT-1 intervenes in the assembly of histones H_3 and H_4, newly synthesised and coupled to the DNA synthesis. For its part, variant $H_{3.3}$ depends for its insertion in the nucleosome on its interaction with another chaperon protein, called HIRA. Its placing, contrary to that of the form H_3-1, is independent of replication.

Faced with the subtlety and the diversity of the changes that can occur at the chromatin level by the possible combinations of the histones at the heart of the nucleosomes, a key question arises. It concerns the epigenetic maintenance of these changes through cellular divisions. The answer rests, according to all probability, in the "semi-conservative" model, such as postulated for the assembly of the nucleosomes. There are reasons to believe that, during replication of the DNA, the nucleosome separates into two half nucleosomes which remain attached to each of the DNA strands present in the replication fork. To take the example of the tetramer $(H_3\text{-}H_4)_2$, this complex would break up into two H_3-H_4 dimers, capable of random distribution on each of the strands of the DNA threads. On contact with each of the parent H_3-H_4 dimers, a tetramer would be reconstituted by matching with a newly synthesised H_3-H_4 dimer. In this way, the parental epigenetic mark would be able to be maintained on each strand of the DNA after replication.

---

3. Kinetochores: The kinetochore is a supramolecular assembly of proteins at the level of the centromeric regions of mitotic chromosomes. There are two kinetochores per centromer being able, in mammals, to interact with 20 to 40 microtubules.

I.7. FROM CONTEMPORAY BIOLOGY TO THE CHALLENGES OF DEVELOPMENT

I.7.1. REFLECTIONS ON CONTEMPORARY BIOLOGY

Thus, and as we have just seen in the first part of this book, after its long "naturalist" period which extended from the end of the 18th century until the middle of the 20th, followed by its attempts to reach the precision and rigour of the laws of physical chemistry (from the start of experimental physiology to enzymology), biology has finally turned to the exploration of the "infinitely small", cellular and molecular. Genes and macromolecules have become, and still are, its favourite objects of study.

However, the life sciences find themselves confronted with new challenges, which are essentially of two types. The first are methodological. The second have to do with the preoccupations of the contemporary world, with the general problems facing our society and our environment. Indeed, on the one hand the question is how we can go further with research, granted the fact that biologists are confronted with the task of organising in a logical (and usable) way the innumerable data provided by molecular analysis. As a matter of fact, since the development of genomics, the researchers (and their databases) keep available a multitude of information about the newly identified genes, their products of expression (nucleic acids and proteins), their elements of regulation, their conformation in space, their modes of interaction, etc. They must therefore move from "molecular descriptions" to real models, to real "interpretative" schemes, compatible with the cellular functions and physiology of living organisms in their integrity. In order to respond to this challenge, contemporary biology tries to establish, with the support of information technology and virtual simulation techniques, new schemes of molecular interaction in complex networks. Furthermore, it must examine to what extent these complex networks allow us to predict the exogenous effects which act on the cells, the tissues or the entire organism. For example, it becomes possible to "model" the functioning of an organ such as the heart, by integrating into this model all the biochemical molecular data, or of others kind, we possess, and to verify its compatibility with what we know about the homeostatic functioning of this organ. This "systems biology" is of an unquestionable interest due to its integrative and predictive approach. Nevertheless, it is worth asking about the risk incurred by this new approach in biology "turning its back" on the true living world, such as it is presented to us in its formidable morphological and behavioural diversity. Is the biology of present times faced with the combined ascendancy of genomics, bioinformatics and models going through an "identity crisis" by becoming too reductive in some way?

In fact, this risk, although legitimate to mention, does not seem to present itself. For example, much of the advances of the molecular biology of the gene and of the genome – and likewise those of developmental biology – have, on the contrary become remarkable analytical tools for more traditional disciplines of biology oriented towards the observation, description and comparative study of living beings. This is particularly true of the research about evolution and taxonomy, but also of that dealing with cellular development, tissue differentiation, physiological function or physiopathology, etc.

Furthermore, at the same time – and that leads us to the other great challenge they face today – biology and molecular genetics have become very important components in modern medicine, pharmacology, agriculture, etc. They often permit us to formulate, in a much more precise fashion than previously, the questions and concrete worries linked to the environment. They should be able to bring new responses to some of these questions. Because what is changing in our present world is a growing perception by society of the fragility of our environment. That is reflected through phenomena as diverse as the appearance of new threats to health (pathogenic agents, pollution, ageing populations), a worrying demographic increase, the dangers of an accelerated urban expansion, the new difficulties in feeding mankind, the effects of climate changes, etc. Hence the questions raised not any more only by the great world decision-makers gathered in international summits, but by society at large: how to ensure a sustainable, economically sound development that is compatible with more equity in the access to the resources of the planet and to healthcare, while avoiding at the same time irreversible damage to the planet's survival? How to conceive the role of sciences in this matter, and particularly that of life sciences?

I.7.2 WHAT CAN SCIENCES DO FOR A SUSTAINABLE DEVELOPMENT? THE ROLE OF BIOLOGY

It is clear that when thinking of the potential "contributors" to the necessary advances for a sustainable development, science must figure prominently. We have already highlighted this. In this way, all the science disciplines are involved, as are the applications which derive from them: mechanics and physics as well as the geosciences, and (in their analytical or synthetic approach) mathematics, computer science, etc., and all these disciplines combined as well as life sciences and biotechnologies. Building decent homes, designing roads, ensuring efficient transport and saving energy represent as many important ways of improving hygiene, alleviating man's isolation or minimizing certain ecological disorders.

However, in this great enterprise, biology and biotechnologies have a specific status and a role whose importance, we think, justifies the present book.

Firstly, because everything affecting the planetary environment, the geosphere, has an effect upon biological equilibrium, man, animals and plants, not forgetting micro-organisms. Secondly, because if we do not take into account essential imperatives such as health, it is very difficult to talk of economic growth. And even worse if hunger is added to the procession of diseases or if fresh water for agriculture will become scarce. Symmetrically, in some way, man, a "thinking being" but also a being following the laws of biology, modifies the biological environment upon which he is dependent. This makes it necessary to have a better knowledge of the environmental equilibrium and to look for new ways of respecting it from now on. Thus, many considerations related to development take us back to biology and its applications, whether medical, agricultural or environmental.

That is why, in the first part of this book, we have tried to explain to the reader, certain aspects of the long, but decisive road to progress of biology. In the following chapter, we will try to show what biology – in its methods and applications – can bring us in our efforts to solve at least some of the major issues involved in the problem of sustainable development. To this aim, we will examine several of the main facets of this problem, such as health and the protection measures to be adopted against its many planetary threats, agriculture and the solutions potentially capable of improving food productivity or the environment envisaged both from the angle of the major energy issues and biodiversity.

However, before entering into the details of these different topics, we will first provide in the following paragraphs a brief insight into them.

I.7.3. HEALTH

Regarding some of the major problems related to human health, such as emerging infectious diseases, genetic diseases, cancer, neurodegenerative diseases, zoonoses and prion diseases, but also, on another level, the socio-economic challenges due to the ageing of populations, it appears that both in the knowledge of the mechanisms of onset of these diseases (whether communicable or not) and in the areas of diagnostics or therapeutics, genetics plays a considerable role today. For example, it contributes to the characterisation of new pathogenic agents or of their modes of transmission, to the determination of the causes of their virulence or to those involved in the host sensitivity or resistance. The sequencing of viral genomes opens up important perspectives to the conception of new vaccines. The possibility of purifying eukaryotic genes by recombinant DNA techniques and of identifying them has revolutionised the nosology of genetic diseases. The biology of a cancerous cell with its list of mutations or epigenetic modifications affecting the "gatekeepers" or sentinel genes (also called suppressors) which can counteract the malignant mutations (oncogenes) has also considerably progressed. Genetics and molecular biology also provide the key to the astonishing phenomenon of programmed cell death (apoptosis), intervening not only in the natural differentiation of our tissues and in our defences against the cancerous cells, but also in some degenerative pathologies. But there is also another field of general human biology that underwent a recent "revolution". We mean the field related to cell development and to reproduction. It benefits from the spectacular advances realized in the isolation and the knowledge of stem cells, a prelude to a real regenerative medicine. The prospects opened up here are considerable. Perhaps tomorrow will we benefit from a real cell therapy covering an increasing number of diseases, including serious neurodegenerative illnesses whose frequency increases with the average age of the population. Finally, the relatively recent discovery of prions, these astonishing proteins capable of adopting abnormal and transmittable conformations without the (apparent) help of genetic material (DNA or RNA), has firmly shaken some of our established concepts of molecular biology by unveiling the mechanisms of a whole series of diseases of men and animals which were hitherto somewhat mysterious (kuru, Creutzfeldt-Jakob, Bovine spongiform encephalitis). The consequences of a psycho-social or economic nature, and also those regarding food hygiene, have been considerable. These discoveries have not only been made possible by the work of epidemiologists and physicians, but also by the so-called "structural" biology, concerned with the crystallographic features of proteins.

It is therefore evident, through these schematic reminders, that there is a growing osmosis between biology, whose most fundamental achievements have been retraced (at least in broad outline) in the first part of this book, and the majority of areas that are related today to the health problems our society has to face.

I.7.4. AGRICULTURE

However, the same remark can be applied to the great sector of agriculture and nutrition. To feed man in the world of tomorrow will demand, as in the preservation of human health, huge efforts, given the difficulties already recalled above such as the rapid population growth, especially in the emerging countries, the scourge of malnutrition and hunger in Africa and South Asia, the pressing threats on the supply of water, the spread of drought, the limitation of arable land, the constraints linked to the safeguard of the environment, etc.

Most certainly, man, from the most ancient time, has had recourse to a variety of procedures to feed himself, and to conserve or transform food; processes which today we would call "biotechnological"! However, the real beginning of modern biotechnology, whether of medical or agricultural inspiration, has without doubt been contemporary of the development of genetic engineering and monoclonal antibodies, followed by the genomic revolution which we have extensively evoked in the first part of this book. Today, in its study and in its applications, plant physiology benefits from the increasingly fine characterisation of what are called genes of interest: those which play a key role in the growth, the robustness, the resistance to biotic (insects, virus) and abiotic stress (drought, temperature, salinity) or even in the synthesis and storage of nutritive elements. Genetic data therefore often make possible the follow-through and the "monitoring" of these principal characteristics of which traditional agriculture can take advantage. However, above all (and there lies the major turning point), there are at present multiple possibilities of developing in plants some properties which will improve their cultivation or utilisation for food or other purposes. These possibilities derive from plant transgenesis. For it is possible to modify the plants by the transfer of one or several genes derived from other plant species or from bacteria. These plants, called "transgenics", of which the best known are those having acquired resistance to insects or drought, are beginning to be cultivated on a non-negligible scale (more than 100 million hectares) in different parts of the world, athough

– especially – in Europe. Some reservations or explicit oppositions are still far from being overcome.

Transgenesis has also been applied to animal breeding, from fish to mammals, but unlike genetically modified plants, at the moment, transgenic animals only benefit from modest interest.

However, transgenesis is not the only application to derive from contemporary biology with regard to the plant kingdom. The recent knowledge of interference mechanisms implemented by small non-coding RNA molecules capable of inhibiting the translation of template RNA into proteins, or to induce its nucleolytic destruction, is in the midst of greatly renewing our conceptions regarding the defence of plants against pathogens. In particular, it is going to lead to the manufacture of synthetic oligonucleotides acting as new agents in the fight against phytoviruses.

I.7.5. Environment – Biodiversity – Evolution

Finally, if there is one area where molecular biology, genomics and bioinformatics have become indispensable auxiliaries in the preservation of our living patrimony and resources (and there almost by definition), it is that of biodiversity. On the one hand, most people are becoming aware of the various threats concerning the diversity and development of the living species (climatic changes, pollution, accelerated urban development, certain practices of deep-sea fishing, etc.). However, on the other hand, faced with the immensity of the challenge represented by the inventory of unknown living species, whether animal or vegetal, and even more that of micro-organisms, the specialists have, there again, a growing recourse to comparative techniques where genetics, and the study of genomes, of their variations and mutations, as well as of the polymorphisms of DNA sequences, and of proteins, from now on will be playing a major role.

Already a few years ago the exhaustive comparative analyses of RNAs present in the ribosomes (these universal organels acting as supports to the biosynthesis of cellular proteins) had allowed Carle Woese to discover a totally unsuspected living kingdom, that of the Archeas, unicellular organisms, intermediaries in virtue of their properties between true bacteria and eukaryotic cells. But the development of the most recent techniques of genomics applied to raw samples of DNA coming, after amplification, from complex "environments", together with the treatment of sequences by information technology procedures – an approach known today as "metagenomics" – <u>will now allow a completely new</u> <u>chapter concerning the</u> <u>world of the infinitely small cellular organisms to be written:</u> bacteria, proto- zoons or single-cell algae, a world of which only a very small part has been catalogued or put in an inventory. New perspectives are thereby open to soil ecology, the study of marine microbiology or that of symbiotic bacteria intervening in intestinal flora.

Molecular biology and genetics are not the only disciplines of biology that we call upon today to explore the diversity of the species and to establish their filiations. <u>Developmental biology</u> (which we have addressed in the first part of this book), a branch of cell physiology which notably explains how the great segmental organisation of the body (i.e., the head, trunk and limbs for the vertebrates, the head, thorax, abdomen, legs and wings for insects, etc.) is put in place under genetic control, brings new complementary insights into the fields of systematics and evolution. Indeed, "it is clear that an essential part of the <u>morphological</u> differences which systematists study, can be understood as resulting from the more or less subtle genetic alterations in the process of putting in place the organisational plan of the species, the determination of the size and the form of the organs, or even the details of their ornamentation and their terminal differentiation" (André Adoutte in "Systematics, to order the diversity of the living", reports RST of the Academy of sciences, n° 11, October 2000, p. 96).

This new field which is opening up by bringing together evolutionary systematics and developmental biology (a field which, in their jargon, the specialists familiarly call "EVO-DEVO") gives a new perspective to the study of biological diversity. We are no longer content to use molecular approa- ches (such as the comparisons of sequences of genes or proteins) as tools permitting the establishment of evolutionary relations among organisms. Rather, we try to characterise the genes capable of providing explanations <u>of</u> <u>morphological changes.</u> In other words, it is not enough to record and clas- sify morphological diversity, it is necessary to have the ambition to explain its genesis.

I.7.6. CONCLUSION

In conclusion, the biological sciences, often in their most fundamental acceptation apparently so remote from any concern for applications, open up in fact numerous perspectives to the knowledge, understanding, and some- times even the solution, of some of the major contemporary problems dealing with sustainable development.

It is true that in many cases, there is still a long way to go between the research and its concrete translation. Most often, as we shall see in the following chapter, biology supplies diagnostic tools, helps to clarify and predict risks, from the area of diseases to that of damage to biodiversity. But in other cases, it already allows us to foresee new solutions (e.g. the protection and improvement of vegetal resources or, on a different level, the fight against certain degenerative diseases more or less linked to ageing, the supply of new sources of energy generating less CO_2, and the manufacturing of new pharmaceuticals better adapted to their molecular target and to the specificities of the individual).

Biology has therefore become a sort of common denominator in the scientific approach to developmental problems. It is however certainly not the only science called upon to solve those problems as they are posed today to a human being confronted with the vagaries of planet earth and of demographic evolution. Biology itself relies on instrumentation techniques, or even methods inspired by other science disciplines. It cannot intervene usefully without being constantly aware of and attentive to social and human sciences, to epidemiology, or to ethics. In short, it can only act in a very large interdisciplinary framework; otherwise it would withdraw into itself and have little exterior impact. However, in this probably lie today its pride as well as its luck: the pride and the luck of offering now a remarkable opportunity for intellectual, technical and strategic convergence, both in the perception of our future development and in accompanying this development in a positive way.

SECOND PART

BIOLOGY AND THE GREAT
DEVELOPMENTAL CHALLENGES

II.1. HEALTH

II.1.1. INFECTIOUS DISEASES (THE REVIVAL OF MICROBIOLOGY,
VACCINES, DIAGNOSIS AND ANTI-VIRAL THERAPY, ZOONOSES,
PRION DISEASES)

II.1.1.1. The return of infectious diseases – Diseases of poverty – Neglected tropical diseases

One of the principal components of the "objectives of the millennium" is to promote a significant decrease in mortality and morbidity that are due to infectious diseases.

During the decades following the Second World War, part of the population cherished the illusion that the spectre of transmissible diseases was becoming less worrying and that, in countries of the North at least, public health should be more concerned with non-transmissible diseases, linked as they are to the new customs of modern life. It is true that from 1960, we witnessed for the following three decades a significant increase in life expectancy which, in spite of disparities, in relative terms, from one region of the world to another, has characterised the developing countries in the same way as the others. Vaccination campaigns on a large scale which had eradicated smallpox and then, almost completely, poliomyelitis, the use of antibiotics and a better awareness of public health problems by international organisations have been, in this respect, essential factors of this positive evolution. Moreover this relative euphoria has led to a slowdown in the teaching of general microbiology.

Yet towards the end of the 1970s, at the WHO (World Health Organisation) and in many other international organisations, the worries regarding infectious diseases would strikingly return in force. On the one hand, the hope placed in using new insecticides in the fight against parasitic diseases (such as malaria) has been somewhat shaken. On the other hand, there did not exist (and still does not exist) any anti-malaria vaccine and the phenomenon of "antigenic variations", particularly frequent in the trypano-somes, meant that many parasites, as well as various viruses, would most likely escape the protection afforded by the vaccines. Finally, and above all, since 1979, nearly 30 new diseases transmissible from animals to humans (zoonoses) have been discovered, diseases including Ebola, HIV/AIDS and new forms of influenza, etc. The AIDS pandemic has caused the terrible losses that we know and which continue to represent, even today, one of the main causes of mortality attributable to an infectious disease, followed by malaria and tuberculosis. These three diseases, sometimes called "diseases of poverty" (WHO) therefore constitute today the priority targets of several international organisations, or NGOs (Non-Governmental Organisations), etc., especially of the international project "DCPP" ("Disease control priori-ties project") which coordinates the efforts and collects the support from the international Fogarty Centre, the National Institute of Health of the United States, the World Bank, WHO, the Population Reference Office and GAVI (Bill and Melinda Gates Foundation).

A general remark is pertinent at this stage. Contrary to the observed increase in average life expectancy observed in the majority of the regions of the world, we are witnessing a significant downturn in sub-Saharan Africa and South Asia, which to a large extent is linked to mortality following AIDS.

According to recent studies conducted by the DCPP, transmissible diseases (infectious and parasitic) are responsible for nearly 60% and 31% of the total problems of public health in sub-Saharan Africa and South Asia, respectively. More precisely, HIV/AIDS, malaria, pneumonia and diarrheal diseases are the predominant infectious diseases in these two regions of the world. Sub-Saharan Africa pays the heaviest toll to HIV/AIDS, malaria, mumps and sexually transmitted diseases. Moreover, it is the only region in the world where infectious diseases constitute the n° 1 factor of mortality, contrary to other regions where the cardiovascular diseases are the main cause. There is, besides, a range of serious diseases in tropical countries which up to now have received very little attention from nations, interna-tional organisations or from large pharmaceutical companies capable of combating them. They are often called "neglected tropical diseases". These must not be confused with the "orphan" diseases which, because of their rarity, hardly provoke therapeutic strategies or concerted research, but which have no prevalence in the South.

These neglected tropical diseases are most often provoked by parasites (protozoons, parasitic worms, bacteria). These are the diseases due to misery and lack of hygiene, plaguing many people in poor countries. The most common include: <u>sleeping sickness, filariasis, onchocerciasis</u> (river blindness), <u>trachoma</u> and <u>leprosy. It is significant that, of the 1,393 new molecules developed between 1975 and 1999, only 16 (a little more than 1%) were destined for these neglected tropical diseases!</u>

However, major political, health and humanitarian authorities in the world are beginning to mobilise themselves as part of the operation entitled *"Drugs for neglected diseases initiative"* (or DNDI) launched by "Médecins sans Frontières". The Gates Foundation, which has already paid nearly a billion dollars in the fight against tropical diseases (including malaria), is precisely in the process of directing part of its efforts towards the neglected diseases. A new scientific review, "PLOS – Neglected Tropical Diseases", was created in 2003 (PLOS: Public Library of Sciences).

The recent declaration by Dr Margaret Chan, director general of the WHO, summarises the contemporary state of mind with regard to the widespread infectious threat and the rise in medical worries: *"population growth, the populating of territories hitherto uninhabited, rapid urbanisation, intensive farming, degradation of the environment, unwise use of antibiotics and other anti-infectious agents, all these factors have shaken up the equilibrium of the microbial world...".*

In this chapter, we will mainly discuss the following themes:

– Microbiology and its revival (the contributions of molecular genetics);
– Vaccinations (new trends);
– Zoonoses and prion diseases;
– Diagnosis and therapy of viral diseases.

II.1.1.2. Microbiology and its revival

• *General considerations*
Before evoking the various signs of the current revival of microbiology, it might be worth briefly recalling what have been the outstanding stages of this science, long considered the "seminal" discipline of biology.

After Louis Pasteur had refuted the theory of "spontaneous generation" and revealed the presence of bacterial germs in the environment, the theory "one disease, one germ" would induce a true blossoming of work leading to the identification of multiple infectious agents: leprosy bacillus (A. Hansen, 1873), gonorrhoea (Neisser, from where the genus "Neisseria" is described, 1879), puerperal fever (E. Roux, 1879), hematozoic malaria (A. Laveran, 1880), typhoid bacillus (K. Ebert, 1880), yellow fever agent (D. Ross and C. Finaly, 1881), tuberculosis bacillus (R. Koch, 1882) and those of diphtheria (E. Klebs, 1882, from where we get the genus "Klebsiella"!), of cholera (R. Koch, 1883), of tetanus (A. Nicolaïes, 1884), of the plague (A. Yersin, 1894), and of syphilis (discovered from spirochetes by F. Schaudinn, 1905). The microbial theory developed by Pasteur, also called "germ theory", would lead to the development of the first modern vaccines (see the chapter on "vaccinology"). As regards Robert Koch, one of the first Nobel prizewinners in Physiology or Medicine (1905), he clearly established the methodological postulates which were behind the isolation of pathogenic germs. These postulates are built on the physical association of the microbe with the disease and its absence in the healthy subject, on the possibility of its isolation from the patient and its cultivation, and on its verification according to which the cultivated microbe injected into a healthy animal causes the disease with its characteristic symptoms, this same microbe becoming isolatable from the ill host. It is also and, above all, as is known, the discovery of the tuberculosis agent, a disease very widespread at the time, which was to establish his reputation and stimulated the prophylactic measures and ways of fighting it (dispensaries, etc.). In 1924, Calmette and Guérin developed BCG, but it was towards the end of the 1940s with the discovery of streptomycin (Waksman, 1952) that this serious illness started to decline for three or four decades before reappearing in different regions of the world (lack of effectiveness of BCG, insalubrity, insufficient recommendation or poorly followed guidelines regarding the use of antibiotics, occurrence of multi-resistant bacilla, etc.).

• *Factors in microbiology revival*

As we have just explained, both because people thought they had more or less succeeded (at least in the countries of the North) in fighting the epidemics of contagious diseases such as tuberculosis, typhoid, etc., and also because of the spectacular progress in molecular and developmental biology, microbiology started to be seen as a "poor parent", especially at the beginning of the 1970's. More or less consciously, the idea prevailed in the public arena that the "inventory" of the major pathogenic microbes was completed and that antibiotics would from then on provide an effective answer to the majority of diseases caused by these pathogens. Biochemistry and molecular biology would generate more vocations among young researchers.

However, it is precisely because of its extraordinary capability of "absorbing", in some way, the other biological disciplines (genomics, structural biology, cellular biology, immunology, etc.) that microbiology has found its vitality again. Indeed, new approaches to the study of bacterial pathogenesis through molecular biology and genetic approaches, with the <u>view of deepening the mechanisms of this pathogenesis and those involved in the defences of infected organisms,</u> are being developed, giving a new stimulus to the discipline.

• *Genomics and virulence*
Genomics is in particular able to provide clear explanations for the <u>virulence</u> (transmissibility, lethality) of certain micro-organisms or certain viruses. For example, we know that the *E. coli* bacterium is in its normal state a beneficial member of the human intestinal flora. Yet, in 1983, a new strain (a new serotype) of *E. coli*, called O 157-H7, was isolated during some outbreaks of diarrhoea in the United States and this strain would turn out to be very virulent. The comparison between the O 157-H7 genes and those derived from the strain of reference *E. coli* K_{12} revealed that, although the two strains have a "common core" of 4.1 million base pairs (probably inherited from an ancestral strain), they nevertheless differed because of the additional presence in the "virulent" version of genomic regions (so-called "islands"). The so-called "O islands" contained nearly 1,400 genes out of the 5,400 bacterial genes. Conversely, 530,000 base pairs of the K_{12} sequence did not exist in O 157-H7 (Perna N.T. *et al.*, *Nature* 409, 529, 2001). Among the 1,400 genes related to the type "O" islands, a number of them were shown to be pathogenic. Another illustration of these perspectives which can derive from comparative genomics resides in the comparative studies undertaken on *Mycobacterium tuberculosis* and *Mycobacterium lepre* (leprosy agent), studies suggesting a strong evolutionary correlation between them, the leprosy bacillus deriving from the Koch bacillus, as revealed by its genome, through the loss of numerous genes of replication.

A striking example of the contributions of molecular genetics regarding the study of virulence can be found in the work recently carried out on the Spanish flu virus. Researchers have been successful in "resuscitating" the 1918 Spanish flu virus, by using the DNA of a victim found frozen in Alaska (*Science*, 7 October, p. 28)! Other teams artificially introduced a few point mutations (a change in the sequence in amino acids) into a protein of the virus, the <u>hemagglutinin,</u> a protein which allows its fixation on glycoprotein receptors present in the host pulmonary cells. They have thus been able to establish <u>that only two amino acid changes in the sequence of the hemagglutinin are sufficient to make the hypervirulent virus of 1918 lose its transmissibility from one</u>

animal (the model in this case was the fox) to another, while at the same time keeping its lethal effect. These two changes lead equally and simultaneously to a significant reduction in the affinity of the modified virus to the predominant viral receptor existing in man (a receptor of type a-2.6, rich in scialic acid and linked to galactose), with on the other hand a parallel increased affinity vis-à-vis the receptor of type a-2.6, characteristic of birds (such as the H5N1 avian flu virus) (T.M. Tumpey *et al.*, *Science*, 315, 655, 2007). The conclusion that can be drawn from these studies is that the extreme pathogenicity of the 1918 virus (which killed 50 million people) was probably due to an exceptional transmissibility linked to the affinity of the influenza virus of that time with a particular receptor of man's pulmonary tissue, a contagiousness which considerably amplified its intrinsic lethal effect.

Numerous pathogenic genes (capsules, toxins, adherence factors, and factors involved in pathogen invasiveness or cell survival) have been cloned, which has made it possible to analyse precisely their role vis-à-vis the target cells. Thus, recent genomics studies have allowed the characterisation of new pathogenic genes which the classical study of viral mutants had not yet been able to bring to the fore. An important concept has gradually emerged from it (Sansonetti), a concept according to which one can observe a frequent topographic "regrouping" of the pathogenic genes in some confined regions of the viral genome called "pathogenicity islands", often present on plasmids[1], bacteriophages or on bacterial chromosomes, which allows and explains the transfer "en bloc" of these virulence effector genes.

• *Target cells and the penetration mechanisms of pathogenic bacteria*
However, it is especially at the level of the target cell itself, a cell damaged and invaded by the pathogenic agent, where significant progress has been realized, contributing largely to the revival of microbiology. This reactivated attention for target cells follows on from research concerning the receptors of pathogen fixation present on the membrane of infected cells (for which we have just given an example with the glycoprotein receptors of the influenza virus).

However, this is equally due to recent work devoted to the penetration mechanisms of certain pathogenic agents (see below) and to innumerable signalisation cascades such as those intervening, for example, in the triggering of innate immunological reactions, bringing into play the "Toll receptors".

_____

1. Plasmid : small circular chromosome of bacteria which reproduces independently of the principal chromosome, which can be transmitted from one bacterium to another and often acts as a "vector" responsible, for example, for a cascade acquisition of antibiotic resistance, among different bacterial species, and which is also used as a tool for gene translation in genetic engineering.

The process by which the target cells, for example the animal cells from an infected tissue, have their cytoplasm invaded by pathogenic microbes has for a long time been mysterious. The study of certain genes belonging to invasive pathogenic bacteria, due in particular to work carried out at the Pasteur Institute by Ph. Sansonetti and P. Cossart, has allowed the precise reconstitution of the principal stages of penetration of various pathogenic species but also of the mechanisms intervening in their transportation in the interior of the infected cell, or in the "passage" from one cell to another.

Two categories in internalisation mechanisms of pathogenic bacteria have been distinguished: one of them is called the type III secretion mechanism. It is implemented, for instance, by bacteria *Salmonella and Shigella*. The other so-called "zipper type mechanism" is used by bacteria such as *Listeria monocytogenes* or *Yersinia pseudotuberculosis*. In the first case, proteins encoded by bacterial genes assemble to form a kind of appendage resembling a syringe, which is used to inject the pathogen into the host cell through a process of translocation, taking advantage of the "polymerisation-depolymerisation" cycle of cytoplasmic actin. *Listeria monocytogenes* (a bacterium which grows in food which has been kept cold for a long time, and which is capable of crossing haemomeningeal, intestinal and foetoplacental barriers, and of surviving by actively dividing inside the phagocytes) penetrates into the cells which it infests by a more complex mechanism. The proteins expressed at its surface, the "internalines", interact with E-cadherin, a receptor of the target cell. This interaction stimulates the recruitment of two other proteins, the α- and β- catenines.

They combine with the actin cytoskeleton of the target cell, making it suitable to facilitate the intracellular penetration of the bacterium or its transition from one cell to another. Many other genes, which have been identified, intervene similarly.

The knowledge of these internalisation phenomena not only throws a totally new light on the mechanisms intervening in the invasive phenomena of pathogenic agents, but may also lead to the production of new vaccines or new synthetic agents capable of blocking, in precise stages, the penetration of the pathogen. However, besides the implementation of these intracellular penetration mechanisms just described, certain bacteria, such as *Yersinia pseudotuberculosis* (an enteropathogenic agent capable of breaking the epithelial barrier of the intestine and having access to lymphatic ganglions), can also grow in an extracellular way by developing a true antiphagocytic strategy. This strategy implies the formation of factors called "Yop", capable of modifying the phagocytosis complex by acting on its cytoskeleton.

Other invasive pathogenic microbes (*Mycobacteria Chlamydia, Salmonella* and *Legionella*) manage to replicate themselves in the interior of the vacuoles of the phagocyte without being carried towards the destructive lysosomes. They synthesise enzymes, also called NRAMP (natural resistance-associated macrophage proteins), which are believed to confer upon them a resistance to the toxic superoxide radicals, produced by the macrophages.

To summarise, in certain pathogenic microbes there is a real arsenal for bypassing the key cellular functions (see the review by G. Tran van Nhieu and P. Cossart, *Medicine Sciences*, 17, n°6-7, p. 701, 2001).

• *Susceptibility genes*

The characterisation of a fairly large number of susceptibility genes present in the infected organism enabled the understanding of the differences in the relative sensitivities of certain ethnic groups regarding a single agent being the cause of pathogenicity. An example of this can be provided by the existence of the human gene NRAMP, whose murin homologue codes for an integral membrane protein, and whose activity is related to the sensitivity of certain populations to the leprosy bacillus (Abel L.)

• *Environment and reservoirs of pathogenic agents*

Modern microbiology, bacterial, viral or parasitic, pays special attention to the influence of the environment as a selection factor favouring the existence of reservoirs of pathogenic agents or their geographical migration. This eco-microbiology concerns nosocomial infections and food toxicology (for example, contamination of milk products by *Listeria monocytogenes),* as well as infections from hotel or industrial air-conditioning systems (as in the well-known Legionnaire's disease, due to *legionella*). Its attention also focuses on the general problem of "accidental" propagation of pathogenic agents, out of their ecological niche, which is becoming a serious burden due to the considerable growth in tourism and frequent plane journeys, etc.

For the microbiologists and epidemiologists, global warming and the changes to the environment caused by man are also factors of preoccupation. The potential impact of climatic changes on vector-borne diseases (for example, those linked to insects) is difficult to predict. However, this change could have, in fact, an important influence on the geographical distribution of endemic areas or on the spread of epidemics (F. Rodhain, in *ECRIN*, 68, 23, 2007). So, for example, a significant geographical spread of the fever virus of the Rift Valley, the Chikungunya virus and of the ovine catarrhal fever virus (Blue tongue) has been observed. Blue tongue reached Corsica and Spain in 2000, probably as a result of global warming. As for the Chikungunya epidemic spread in 2005 in Reunion Island, this was seemingly produced as a result of insecticide treatments which were too intensive and which modified the biological equilibrium among various mosquitoes. This would have led to the replacement of *Aedes aegyptii*, a carrier of malaria, by *Aedes Albopictus*, a transmission agent of the Chikungunya virus. Regarding the Rift Valley virus, responsible for deadly epidemics in Egypt, its increased presence would have been related to the construction of the celebrated Aswan dam and the ensuing water reservoirs. The recent increase in parasitic diseases, and especially leishmaniosis in the Dakar area, can be explained by a similar set of circumstances: leishmaniosis would have been caused by the work leading to the creation of a new dam on the Senegal River (A. Capron, personal communication). Another recent study also dealing with leishmaniosis, but in North East Colombia, has shown that its incidence was very sensitive to the changes in temperature due to the "El Niño" phenomenon. Forward-looking studies on the possible introduction, in France, of new animal diseases or the increase in their incidence (visceral leishmaniosis, horse plague, blue tongue, etc.) as a result of climate changes, are being led by groups of experts in conjunction with the AFSSA (the French Agency for Food Safety).

Thus, a real environmental microbiology involving, in addition to biologists, virologists and physicians, some epidemiologists, entomologists, etc., is beginning to predominate.

Within the same framework fall other changes in human behaviour, resulting from interactions between humans and animals from tropical regions, where reservoirs of animals carrying viruses that are transmissible to man are often found. Such is probably the origin of AIDS.

However, other considerations and observations also are at the origin of the microbiological revival. This is the recognition of the role played by pathogens of viral nature in the occurrence of cancers[2] (for example the HBV virus) and the occurrence of primary hepatocarcinomas, but also the close relationship which has been described between the cervical presence of papillomaviruses and the development of cancers, which has warranted the Nobel Prize in Physiology or Medicine for the German virologist Harald zur Hausen (2008). Another illustration is found in the action of the bacterium *Helicobacter pylori* in the triggering of gastric ulcers. In fact, it would appear that transmissible, pathogenic agents (sensitive to immune reactions or to antibiotics) can be responsible for the occurrence of "non-communicable"

---

2. The awarding to doctor Zur Hansen, of the 2008 Nobel Prize in Physiology or Medicine, for having shown the role of papillomaviruses in cervical cancers, is a good illustration of this point.

illnesses, which, so far, were thought to be totally independent of infectious diseases[3].

II.1.1.3 Vaccinology

• *Historical insights and generalities*
   The concept of vaccination was born in China. Infants were injected with the contents of pustules of smallpox from patients who had developed an attenuated form of this disease with the aim of protecting them.

   The first systematic intervention (1796) to protect infants against this disease, which was extremely rife in Europe, was due to an English doctor, Edward Jenner. He had the premonitory idea of giving a child a form of smallpox that is benign to man, namely, <u>Vaccinia</u>, a disease affecting cows, by taking a sample of the liquid from a Vaccinia pustule. He successfully managed to protect the child from the virulent form (the name "vaccine" therefore comes from this process since it evokes the term "vacca" which is the Latin for "cow").

   However, it is Louis Pasteur towards the end of the 19th century to whom we are indebted for the considerable enlarging of this prevention principle and for having extended <u>the practice of experimental vaccination</u> as well as its clinical application.

   In fact, by discovering some procedures leading to the attenuation of germs, he was able to put his observation into practice by developing the first vaccine against cholera in hens, then to extend his experiments to red swine fever, anthrax and finally to rabies, a disease which was devastating and frequent in Europe at the time[4].

   The principle of protection against an infectious disease by administrating the corresponding agent in its attenuated form would be extended thereafter to the production of numerous other vaccines, such as those established against diphtheria, tetanus, pertussis, tuberculosis and yellow fever, just to mention a few.

3.Besides the well-established case of causal relationships that can occur with a significant frequency between infection by HBV and hepatocarcinomas, it has been reported that the hepatitis C virus or even the *Helicobacter pylori* or *Campylobacter jejuni* may be implicated in the occurrence of lymphoma.
   4. In the period after the war, the Pasteur Institute would also play a major role in the development of a vaccine against poliomyelitis (1954), then against hepatitis B (1985).

• *Different types of vaccination*

Very many vaccines, which could be qualified as first-generation vaccines, because they were conceived according to the procedures developed by Louis Pasteur, are "live attenuated vaccines". This is the case, for example, of vaccines intended to protect against measles, poliomyelitis, tuberculosis, etc. To cause this attenuation, one often has recourse to the isolation of the "mutant" forms of the pathogen (bacterium or virus). The attenuated germ retains its replication properties in the person being treated, while losing its virulence.

One can also make use of the invalidation technique based on homologous recombination (gene knock-out) to free a pathogenic agent from its virulent germs. Often, one proceeds to the elimination of several pathogenic genes at the same time or one isolates multiple mutants to avoid an eventual "return" to virulence. Yet, security is imperative; despite implemented precautions, the fear of seeing, even exceptionally, the reacquisition of virulence by a live vaccine has led people, over these last years, to favour the use of the so-called "molecular vaccines". Such vaccines are constituted by "purified antigens", whose immunogenic and protective powers are carefully checked before use. For guaranteeing an absolute purity, vaccinating antigens are often produced by genetic engineering. As an illustration, one can mention the utilisation of the antigen Hbs, obtained by cloning in Chinese hamster cells (P. Tiollais) or in yeast (Merck process), as an efficient, highly purified vaccine against type B hepatitis, a disease of rather high prevalence in some emerging countries in the South and which can degenerate into hepatocarcinoma. The use of the rabies virus glycoprotein as an efficient vaccination procedure, also purified by cloning techniques, constitutes another example.

However, it is not always easy to characterise the antigens endowed with vaccinal properties from the multiple molecules intervening in the constitution of micro-organisms. Often, research is carried out on the well-defined molecular entities present on their surface, as they should be much more easily accessible to the antibodies. The knowledge issued from the identification of genomic sequences can prove very useful for the inventory of potential antigens, particularly if this identification leads to the clarification of the nature of virulence genes.

However, it is equally important to determine the "domains" of these antigenic proteins which are recognised by the T lymphocytes (cytotoxic or NK), domains called "**epitopes**". These epitopes are "small-sized peptide" sequences which, after association with molecules of the major histocompatibility complex, can be "presented" to the receptors of the lymphocytes. Some procedures allowing the definition and optimisation of sequences of amino acids "spotted" as potential epitopes for a given HLA group do exist and are implemented in particular by the Genopoles in liaison with industry.

One frequently tries to amplify the immune response to the vaccinating antigens by appropriate molecules which function as activators of the dendritic[5] cells (at the level of receptors of Toll type – cf. the work of Jules Hoffman), which leads to the production of various mediators (cytokines, chemokines) playing a role in the specific immune response.

Other routes utilised to elicit vaccination, such as the oral method, are used. This type of vaccination can involve, besides the classic immune response, a mucosal immunity (e.g. IgA produced by the intestine, the saliva, the vagina, etc.). Equally, an intranasal method has been developed to vaccinate against influenza.

One of the current trends of vaccinology concerns therapeutic vaccination, the rationale of which is to stimulate the immune system of (already) sick patients. It enables one to control chronic viral infections, due to human oncogenic papillomaviruses (herpes-like virus, HPV16 and 181) associated with cervical cancer or with the HBV or HCV viruses.

It also seems appropriate to mention at this point another type of vaccine, of potential clinical use, but whose effects for now are essentially being studied at the experimental level: these are "DNA vaccines". Here, one no longer uses viral proteins for chronic viral infections, but the DNA that is encoding these proteins. The DNA (or portion of the DNA) coding for these proteins is generally inserted into a bacterial plasmid, linked with a eukaryotic promoter (to allow expression in a eukaryotic context, *in vivo* or *in vitro*). It has been observed that immunisation by the DNA method against different viruses (herpes, hepatitis B and C, cytomegalovirus, rabies, influenza) induces together the production of neutralising antibodies and a cytotoxic cellular response. Some clinical tests are underway, but in humans their results are still inconclusive.

From this very schematic survey, we can see that either by the intervention of attenuated forms of the pathogens, or by recourse to molecular techniques (purified antigens), we have available numerous vaccines capable

---

5. Dendritic cells, present under the skin, in the muscles, the blood etc. initiate the responses of the T cells by their capability of "presenting" the antigenic peptides (in association with the CMH complex and immuno-stimulating molecules), to cellular receptors of the T lymphocytes, leading to the activation of the latter and to the outbreak of effector responses.

of protecting us against numerous viral affections, such as: measles, rubella, mumps, hepatitis, rotaviruses (responsible for severe childhood diarrhoea), papillomaviruses, etc. Furthermore, it is possible today to have one vaccination against 6 distinct diseases in a single shot.

• *The challenges posed by AIDS, malaria and tuberculosis*
    However, despite much research (impossible to describe in detail here), we still do not have vaccines available against two of the most formidable pathogenic agents: AIDS and malaria. Although the AIDS virus was discovered in 1983 by Luc Montagnier, Françoise Barré-Sinoussi (2008, Nobel prizewinners in Physiology or Medicine) and colleagues at the Pasteur institute, its very high rate of mutability makes the task difficult. According to some microbiologists, it would seem that for an effective vaccine to be achieved, it would be necessary to induce together a "cellular" response (T lymphocytes)[6] as well as a humoral type response (producing neutralising antibodies). Remember that, according to the figures published by the WHO and the UN/AIDS in 2006, it was estimated that 39.5 million people in the world are infected with HIV and that nearly three million died in the year 2008 as a result of the disease. Regarding malaria, numerous researchers worldwide are waiting for "candidate vaccines", i.e. for molecules or molecular complexes isolated from *Plasmodium falciparum* assuring, after administration, a durable immune response. Work is being conducted by different institutes in this regard, notably in Bamako (Mali) at the heart of the Malaria Research Training Centre (MRIC), as well as in France at the Pasteur Institute and in many other regions of the world. The "European Malaria Vaccine Initiative" (from the European Union) is supporting various projects (some in conjunction with the Pasteur Institute) aiming to identify candidate vaccines against malaria: it involves, for instance, some trials carried out with a molecule existing on the surface of the parasite during its cycle of erythrocytic development (the molecule MSP3) and another one (LSA) at the hepatocytic stage of the parasite. As for tuberculosis, we know today that BCG has lost its effectiveness for a large part of the population affected by the scourge.

    More generally, it can be said, in retrospect, that the vaccination programme of the WHO launched in 1974 has unquestionably saved a large number of human lives. Nevertheless, according to this organisation, one can estimate, at the start of the 21st century, that 37 million infants in the world have not been systematically vaccinated.

---

6. Peter Doherty and Rolf Zinkernagel were awarded the Nobel Prize (1996) for their cardinal discoveries concerning the spicificity of the cell-mediated immune defence.

Particularly worrying is the threat which AIDS poses during mother-child transmission and also that which is due to the "neglected" diseases. These are diseases for which the prevention cannot be ensured (either because vaccines do not exist or because they are not available in the endemic countries). Nor are efficient treatments accessible, due to poor economic conditions. Research and development concerning these diseases remain insufficient. They include numerous respiratory illnesses, diarrhoeal diseases, parasitoses, etc. However, some industrialists have recently undertaken to act against these neglected diseases despite the feeble "return on investment". The Bill and Melinda Gates Foundation (with an endowment of 25 billion euros) dedicated to the health of children in developing countries (and in particular to the search for new vaccines) has included the development of certain anti-diarrhoeal vaccines in its priority programme.

It is clear, however, that health policies in developing countries cannot rely solely on vaccination campaigns. They should also rely on a set of measures, including hygiene, prevention and education. The role of women is dominant in propagating such measures at the level of the family unit, as advocated strongly by the World Health Education Programme.

II.1.1.4. Zoonoses

Table 2.
For detailed comments, see the chapter on emerging diseases.

Some examples of emerging viral diseases[7]		
Type of viral disease	**Consequences**	**Transmission etc.**
HIV/AIDS	About 40 million people throughout the world live with HIV infection (2006)	2.9 million deaths (in 2006)
SARS	Severe Acute Respiratory Syndrome – epidemic started in China (2002)	800 deaths (2003-2004)
EBOLA	Identified for the first time in the Democratic Republic of the Congo (1976)	Extremely virulent, high level of mortality (85%) but effects limited

---

7. Usually of animal origin.

CHIKUNGUNYA	Re-emergence of an epidemic on La Réunion and Mayotte (2005-2006)	Highly contagious (38% of the inhabitants of La Réunion. Epidemic under control (2007-2008)
HONG-KONG FLU	Epidemic 1968	2 million deaths
AVIAN FLU	Virus H5N1 – Transmission avian. Not contagious between humans	279 human cases recorded (2007) 169 deaths
MARBURG	Origin: Ugandas apes	Highly virulent, geographically under control
WEST-NILE	Origin: avian. Wide geographic distribution	12000 people contaminated in the USA (2003-2004)

Anthropozoonoses (abbreviated to Zoonoses) are infectious or parasitic diseases, naturally transmitted by animal vertebrates to humans (and vice versa). More than 200 of them have been identified to date (arboviruses, brucellosis, echinococcosis, rabies, tuberculosis, etc.). Man pays a heavy price, especially in the developing countries (J. Blancou and P. C. Lefèvre, 2007).

As the National Academy of Medicine and the veterinary Academy of France emphasized in a common declaration in March 2006: "it is more evident than ever that to control zoonoses, it is the animal reservoir which must be tackled as a priority. This fight will only have a chance of success if there is cooperation between physicians and veterinarians on the one hand, and between the developed countries and the developing countries on the other". Zoonoses, true emerging diseases of these last decades, are very often due to viruses. The alarms raised by them are very strong and amply justified.

• *AIDS (the HIV virus)*

Consider the AIDS virus, whose reservoir was formed by some great apes of tropical Africa. Research carried out in Gabon in 1989 by scientists from the IRD (Institut de Recherche pour le Developpement) had established that a domestic chimpanzee was the carrier of a virus type related to HIV-1, an observation that had led to the thought that the chimpanzee could be the natural reservoir of the virus. However, taking into account the isolated character of the observation, doubt remained. Further research was therefore pursued, particularly in an attempt to identify the nature of viruses present in the droppings of monkeys and gorillas. In wild chimpanzees, these investigations succeeded in showing the presence of significant quantities of HIV 1. More specifically, the sub-species, *Pan Troglodytes troglodytes* (Congo basin) is infested by the HIV-1 virus belonging to groups M and N, while group O has been discovered in gorillas (IRD, activity report 2006, p. 25). It is thought that contamination of humans would result from hunting accidents or the consumption of monkey meat (which would have taken place in the 1940s). The simian retroviruses are very widespread in central Africa. Human / monkey contact is amplified by the massive deforestation. The AIDS pandemic is a true global health disaster. Recall the figures: nearly 40 million people living with HIV in 2006 (WHO) and 2.9 million deaths in the same year, sub-Saharan Africa having the highest prevalence rate (with 34% of the population contaminated in some countries). The mobilisation against this health crisis was more intense at the international level than for any other epidemic; "the AIDS pandemic has focused in twenty years the social upheaval that other epidemics have not elicited" (A. Desclaux, *journal du CNRS*, 208, p. 23, 2007). Although di- or tritherapies have significantly extended life expectancy (when the people have access to them), one is still looking, as mentioned before, for an effective vaccine (see the vaccinology chapter).

• *SARS*

Most often the "emergent" pathogens (some of which owe this character of "emergence" to the fact of having being subjected to recent mutations) appeared in the population as a result of contact with the animals serving as primary hosts. This is why environmental disturbances (such as deforestation, the cultivation of previously wild land and the eradication of predators, as well as the illicit trade in wild animals) can become very significant factors in the sudden outbreak of epidemics. With the occurrence of supplementary mutations, the Zoonoses can next become transmissible from man to man.

Thus, at the end of 2002, appeared the severe respiratory syndrome virus (SARS). The epidemic started in the Guandgong region in China, spread through Asia, then reached other continents (tourism, travelling of medical teams)[8]. The animal source appears to have been a small mammal eaten for its meat, the civet musk. The SARS epidemic has caused around 800 deaths. The agent responsible was identified in 2003 by teams from the Pasteur Institute at the request of the WHO and by Dr. A. Osterhaus, a virologist working at the ERASMUS centre in Rotterdam. The disease is due to a coronavirus of a new

---

8. It was a physician, an asymptomatic carrier of the virus, who contaminated 12 people in a Hong Kong hotel and these, in returning to Canada, Ireland, Vietnam and Singapore, propagated the virus across the planet.

type. Ultimately, the epidemic appears to have been controlled by quarantining the patients.

• *Ebola*

Another example of an emerging viral disease from an animal source is supplied by the hemorrhagic fever known by the name of Ebola. The virus was identified for the first time in the Democratic Republic of Congo in 1976 when it was shown to be an extremely virulent pathogenic agent but with only limited effects. Thereafter it reappeared in South Africa, Gabon and Congo Brazaville. Its original habitat seems to be the bat. The virus causes a mortality rate which could reach 85%. It is transmitted to humans by contact with dead or sick primates. The management of these crises is sometimes made difficult by the (justified) "terror" inspired by the disease among local populations.

• *Chikungunya*

Sometimes, the inter-human transmission of the zoonose is not systematic, i.e. it does not always depend on a direct transmission from a patient to a healthy person. It may depend on a vector. This is what has been recently observed at the time of the epidemic caused by the Chikungunya virus in the Reunion Island and Mayotte. It seems that the virus strain, of African origin, became adapted by mutations to a vector, the mosquito *Aedes albopictus*. The disease which, before the epidemic of 2005-2006, was considered somewhat benign, has been accompanied by serious symptoms which may be related to the mutations which had simultaneously modified the habitat of the virus. The contamination rate has been estimated to be around 20,000 people per week, 38% of the population of the island being affected. The epidemic dried up at the end of 2006, probably due to systematic use of insecticide agents and because of immunity acquired by contamination. However, some fear a resurgence of the crisis, that could result from the resistance of the mosquito vector to the insecticides.

• *Avian flu*

There is another alarm, that of avian flu caused by the H5N1 virus (see above).

This disease, as its name indicates, affects birds especially, but can also be transmitted to humans. In May 2007, there were 279 human cases, spread through 11 countries, and 169 deaths. At present, it seems that the transmission from bird to man, although real, remains moderate and that inter-human contamination has not been reported. Fears remain strong and the confinement of birds and even preventive culls are justified by the concern that mutations could change the transmissibility of the virus, creating a catastrophic pandemic. Let us recall that in 1918 (see above), Spanish flu caused nearly 40 million deaths! In 1957, Asian flu had 4 million victims and in 1968, Hong Kong flu had 2 million. It is worthwhile remarking that the HN51 virus generally multiplies in birds, therefore in animals whose temperature is close to 40 °C; while the temperature of the human respiratory tract is close to 33 °C. Therefore, viruses infecting humans are, in all likelihood, viruses having undergone a mutation in their replication system. Generally, the virus spreads to the deep respiratory tract, the upper airways not containing adequate receptors: it probably explains why, as it cannot be expectorated by coughing, inter-human transmission is difficult. Unfortunately, if after mutation HN51 were able to colonise the upper respiratory tract, a veritable pandemic would follow!

• *Marburg, West Nile*
    Other emergent viral diseases are equally formidable even though they are less well known in Europe. For example, the diseases caused by the West Nile virus (the "monkey pox"), the Marburg virus or the hemorrhagic fever with renal syndrome.

    Without doubt, the West Nile virus is the most worrying in this series. It was discovered in Uganda at the end of the 30s, but it is widespread across different continents. It is transmitted from infected birds to humans by the intermediary of mosquitoes. In the years 2003 and 2004, an epidemic struck the United States (14,000 people contaminated, 600 deaths). There have only been rare cases in Europe.
    The Marburg virus also comes from Ugandan monkeys, themselves infected from an animal source unknown to this day. It is transmitted to humans by animal droppings and leads to death in the majority of cases.

• *Recent studies*
    Research on zoonoses is actively pursued in the world. In particular, it concerns the identification of the responsible agents, most often of a viral nature, and the study of their characteristics, especially those related to their genomic features. Thus, significant progress has been made in the genomics of the most dangerous viruses, such as Ebola and Marburg (H. D. Klenk), in the understanding of their replication mechanisms and of the role of the cells specialised (macrophages, monocytes, endothelial cells) in their pathogenesis (hemorrhagic fevers).

    The Erasmus Medical Centre in Rotterdam, led by Dr. A. Osterhaus, has been at the forefront in identifying many zoonotic viruses: for example, the new influenza virus discovered in Hong Kong (1997), the human avian flu virus type H5N1 (1997), the human metapneumovirus (2001), the SARS coronavirus (2003), the human avian flu virus type H7N7 (2003) and a new human coronavirus HCOV-NL (2004), as well as various types of viruses responsible for zoonoses among different animals (birds, sea-lions, monkeys, etc.). The Pasteur Institute has equally played a very important role in the phase of characterisation of the viruses responsible for zoonoses, as well as in the elaboration of new vaccines and in the study of the commune defence mechanisms related to the host genetic background.

Regarding **the severe acute respiratory syndrome** (SARS), various studies have focused on the characterisation of the host factors which influence the pathogenicity of the SARS-COV virus in different animal species. The locations of the receptor of this virus, ACE_2, and of the viral antigen in the ferret, cat and macaque have shown remarkable differences in its distribution. "Transcriptional profiling" (using DNA chips) of the pulmonary tissues of the infected macaque have brought to light the existence of an "innate" type of response characterised by the activation of different genes coding for interferons, a response accompanied by the induction of elevated levels of the interleukins Il-6, IL-8 and IP-10; which is consistent with the deep acute respiratory dysfunction such as was observed in humans after infection.

The immuno-histochemical studies have shown a pronounced inhibition and translocation of the phosphorylation of STAT-1 in cells infected by SARS-COV, which explains how the virus manages to escape from an immunological early type response.

Other aspects of pathogenesis, accompanying the infections by viruses responsible for zoonoses, which have benefited greatly from recent advances in structural biology, concern the properties and role of the hemagglutinins (HA) of the influenza virus. It is indeed a key antigen of this type of virus, intervening predominantly in the attachment of the pathogenic agent to the glycoprotein receptor of the infected cell, as well as in the phenomenon of antigenic variation (allowing the virus to bypass the host defences) and in the process of membrane fusion.

Very intensive studies due, chiefly, to the laboratory of John J. Skehel (National Institute for Medical Research, United Kingdom) and Don Wiley (Harvard Medical School), studies initiated in 1981 by the crystallisation of hemagglutinin, have in fact been carried out at the atomic level. Unexpected subtleties were observed in the specificity which is proper to the different modes of attachment and recognition harboured by the multiple forms of the influenza virus, as well as in their mechanisms of fusion with the cellular membrane of the infected host.

Indeed, regarding the fusion of hemagglutinin with the membrane of the host cell, it was shown that it involved a preliminary cleavage of this viral glycoprotein by a cellular protease, a process necessary to the infectivity of the virus. Influenza "recombinant" viruses presenting in their hemagglutinin a "modified" cleavage site are at present the object of attention as potential vaccines (Klenck).

II.1.1.5. Diagnosis and therapy of viral diseases – An overview

The abundance of viral diseases requires, increasingly, techniques for rapid diagnosis. These may vary in nature, but can attain a high degree of sophistication.

The traditional methods that continue to be widely used are indirect ones, in that they involve the detection of specific antibodies against the virus in samples from the patient. For example, all clinical laboratories responsible for biological testing of donated blood characterise, through appropriate batteries of tests, antibodies directed against two particularly harmful viruses, HIV (human immunodeficiency virus) and HCV (hepatitis C virus).

In some laboratories, in order to detect the early presence of these two viruses in the blood serum, one uses a process that permits direct quantification of the viral RNA, or of specific viral antigens (identified by batteries of monoclonal antibodies on a solid support).

For other classical methods, for example when diagnosing a viral infection in a hospital, one often has recourse to techniques such as direct cell cultivation, electronic microscopy examination or immunoenzymatic testing. The disadvantage of these methods is in their relative slowness in providing results and in the fact that the culture of viruses for diagnosis purposes is fairly ponderous. Recently, molecular biology has allowed the development of rapid diagnosis tests, often quantitative and automatic. These tests are based, for example, on the molecular hybridisation of probes labelled with the genetic material of the virus or on an amplification of the viral genes by the PCR technique (Polymerase chain-reaction). Most often, one has recourse to probes detecting fluorescently-labelled viral RNA or DNA, after amplifying the viral material to be tested; this operation is achieved by using biochips. As already described, these systems consist of solid supports covered by from several hundred up to several thousand DNA probes of different but well-known sequences. The point of interest of this technique is that it allows the simultaneous detection of several pathogenic viruses in the same sample (multiplexing). Due to the selective choice of probes of known sequences corresponding to genomes from different strains (but from the same viral species), one can equally clarify the genetic nature of the infectious strain (genotyping).

Finally, alongside the techniques of classical gene amplification (type PCR) (where the detection of the amplified product only takes place at a stage following its application on a support containing the specific probes) there are today techniques called real-time amplification techniques: the detection is made at the same time as the amplification, the measurement of the fluorescence being effectuated at each cycle catalysed by the polymerase. In these recent techniques, the duration of the amplification is reduced (use of capillaries and air flux). Although these technical improvements, applied to the diagnosis and direct molecular typing of the viral agents, are beginning to be the subject of active development by different industrial firms, the research into antibodies, that is to say, the indirect detection of the viruses, still keeps its serious advantages from the fact of its robustness and its low cost!

•*Antiviral therapies*
The fight against viruses is not limited to the diagnoses of the diseases they provoke. Unfortunately, as far as antiviral therapies are concerned, antibiotics are inefficient. Most often, one has recourse to synthetic chemical agents, and in rare cases to monoclonal antibodies or even to passive immunotherapy.

Without pretending to offer a survey of the multiple examples of production of antiviral agents, we will remind here the strategies which seem to inspire their development, strategies which imply a deep knowledge of proteins or of enzymes specific to the target virus.

On the chemical side, we must remember that combinatorial chemistry is often the starting point for research. This technique is used to establish vary large families of chemical components (combinatorial library), either by chance or by a rational approach; for example, by "transplanting", in some way, a multitude of "active groups" onto defined chemical skeletons, in such a way as to obtain thousands or more chemical "variants". After which, these molecular collections are screened on targets of interest, most often proteins derived from the pathogenic agent or whole cells of micro-organisms. However, this approach, long and costly, can only be implemented by big pharmaceutical firms which have at their disposal large budgets and a significant number of research staff. Rational screening is not always at the origin of active molecules used on a very large scale (see, for instance, the discovery of AZT, of Viagra, etc.)

In order to "rationalise" the combinatorial chemistry and automated screening approach to a specific target, one makes use of computer simulation or data analysis in order to "design" the active molecule (rational drug design). For example, one will attempt to design a molecule which is a candidate for being used as a medical drug, by exploiting its (potential) similarity to the chemical structure of a natural ligand of a receptor, with the aim of blocking the receptor in question (synthetic chemical analogue).

A particularly effective variant of "molecular modelling", known as "de novo drug design", uses the delimitation in space of groups of atoms in the candidate molecule (e.g. an enzyme inhibitor) – groups called "Pharmacophores". Several of these groupings form a three-dimensional motif capable of acting, by complementarity, with the analogous structural characteristics of the "active site" of the enzyme. Of course, it is understood that the relative distances of the pharmacophore groupings of the candidate molecule must be calculated with extreme precision. This technique of computer design of active chemical agents has met with great success. For example, the **Cozaar**, a regulator of arterial tension, acting on the renin angiotensin biochemical couple, has been conceived in this manner. However, what concerns us more directly in this chapter is that it is particularly significant that the protease inhibitor of the HIV-1 virus, one of the major components of the AIDS tritherapy, today implemented on a large scale, was conceived in the same way.

We imagine that these new approaches to the identification of molecules endowed with antiviral activity often imply a precise molecular knowledge of viral enzymes. Besides the example of the HIV protease quoted above, we can also mention the research carried out (J. Gutenberg) on the enzymes which play a role in the replication of the hepatitis C virus (NS5B-ARN polymerase, NS3 protease, NS3 helicase). Another strategy of the fight against viruses lies in the use of the "antisense RNA". Generally, synthetic antisense oligonucleotides are produced with a chemical feature that makes them capable of forming hybrids with a complementary sequence in the RNA of the virus. Most often, the result is an inhibition of viral RNA translation into proteins, a blockade of splicing or of RNA replication (if it is a RNA virus), etc. This strategy is used to impede the action of different virus families (retroviruses, herpes, papillomaviruses, etc.). Various research projects have been carried out with antisense oligonucleotides, research aiming to inhibit the synthesis of major proteins in the development of HIV (Greg, Rev, Tat). Several anti-HIV oligonucleotides are in clinical trials.

Numerous experimental studies have also been carried out according to the same strategy as in the case of the hepadnavirus, whose prototype is HBY, the hepatitis B virus; they have shown experimentally that it is possible to block certain key stages of the translation and the replication, as well as the encapsidation of this virus. *In vivo*, certain oligo-antisenses have proven very active in their capacity to prevent the replication of the avian hepatitis B viruses at the level of the hepatocytes.

Very convincing results have been obtained clinically in the utilisation of these antisense RNAs for inhibiting the cytomegalovirus (CMV) which causes retinitis in patients suffering from AIDS.

II.1.1.6. Prion diseases

The discovery of prions and the diseases to which they lead, has represented a true revolution. It was manifested at the scientific and medical levels, but also in the area of food security, and by its socio-economic and ethical consequences. It has led to new sanitary rules, both national and international[9].

• *The "prion" protein PrPsc*
We know today that prion diseases are caused, not by viruses or bacteria, but by a protein which is adopting an abnormal conformation, called PrPsc. This protein exists in its natural state (PrP) in the brain and probably intervenes in the sleep process. The PrP protein can undergo a change into PrPsc and the interaction of the abnormal form with the natural form causes, in the latter, a change in conformation, so that it becomes capable in its turn of transferring the anomaly to other PrP proteins, and so on. The accumulation, without doubt very slow, of PrPsc proteins in the brain leads to serious neurological problems. The three-dimensional structure of the PrP protein and the different prion-like proteins capable of causing pathological effects was established by a Swiss researcher, Kurt Wütrich (Nobel prizewinner in 2002), but it is the American biologist, Stanley Prusiner (Nobel prizewinner in 1997) who first purified the infectious agent responsible for one of the long-known animal zoonoses, "scrapie". In fact, he was able to demonstrate that this disease was caused by a thermostable protein, not associated with nucleic acids, a protein which he called "prion" (for Protein Infectious) and for which he described the mode of action.

---

9. See "How have the cows become mad?" Maxime Schwartz, ed. O. Jacob (2001).

That a protein could be capable of directly transmitting some kind of physico-chemical "information" to another one, without the intervention of a nucleic acid (up to that time considered as the only chemical support of genetic information), and that it could propagate itself as an infectious agent, was contrary to all the dogmas of molecular biology. However, in 1966, the radiobiologist Tikvah Alper had observed that the "agent" responsible for "scrapie" was resistant to considerable doses of UV radiation, capable of inactivating even the smallest known viruses.

In 1967, shortly after the publications of Alper and Latarjet, John Griffith developed in his journal *Nature* a hypothesis according to which a (sole) protein could be capable of reproducing and being infectious at the same time. This work challenged, for the first time, the viral nature of the pathogenic agent. However, some authors, while admitting the key role of the PrP-PsC protein, questioned that a protein could, by itself, constitute an infectious agent (that is to say, that it was capable of perpetuating itself in the tissues and eventually becoming transmissible). The existence of "minuscule viruses" accompanying the protein agent or of discrete "cofactors" has been advanced. Thus, Prusiner himself proposed that a cofactor, the X protein, would play a role in the "chain" reaction leading to the "distortion" of the set of normal PrP proteins.

Whatever it may be, the fact that the pathogenicity of the prions results from a process of trans-conformational "recruitment" (or if one prefers, from propagated conformational changes) is clearly established. In fact, direct injection of the infectious form, PrPsC, into mice "deprived" of the normal protein-prion PrP – as a result of the invalidation of the normal PRNP gene ("knock-out" experiment) – does not lead to the disease! This is proof that it is linked to the formation of an interaction between the abnormal protein and the naturally occurring one. This kind of interacting cascade, initiated by contagion from one protein to another, does not exist only in mammals. It can be observed in yeast, in certain conditions. Thus, it brings into play an infectious agent, PSP+, a derivative of a normal sup 35 protein with which it associates.

The normal prion protein is a glycoprotein. It comprises around 250 amino acids and is present in all mammals[10]. Although expressing especially in the brain, it is also found to be present in the immune system (dendritic

---

10. The protein in its natural state possesses a long mobile extremity (tail) attached to the spherical (globular) domain of the molecule, comprising three segments coiled into a-helices and one element coiled into a beta sheet. The skeleton of the modified (infectious) protein would be deprived of helices (see Lledo, P. M.: "Diseases from prions" What do I know? Edit. PUF (2002)).

cells) and in the digestive tract (the form PrPsC is also detected in different lymphoid organs: tonsils, spleen, Peyer plates).

Regarding the role of the normal PrP protein, we are reduced to hypotheses: protective effect against early apoptosis, intervention in the elongation of the axons of nerve cells, intracellular transport of copper, agent of anti-oxidative stress. Other authors (Baumann *et al.*, 2006) postulate a protective effect of myelin. The pathological prion, the cause of neurodegenerative diseases, can be detected, in the first stage, at the level of the immune system and, as has already been indicated, it is especially attached to the dendritic follicle cells which are present in the lymph nodes.

According to Adriano Aguzzi of the University of Zurich, it would then spread to the peripheral nerves before reaching the brain. It has indeed been shown that PrPSc could be absorbed by the presynaptic endings of neurons (J. G. Fournier, 2001).

According to other authors, the blood system would constitute a different access path to the brain. Some see the mesenteric plexus of intestinal tissue as playing a role.

At the level of brain tissue, it is generally accepted that neurodegenerative disorders are preceded by an attack of microglia p-cells which would release neurotoxic substances. It has been possible to obtain transgenic mice in which the microglial system is somewhat "paralysed" and which are resistant to various neurodegenerative diseases such as encephalomyelitis (F. L. Heppner *et al.*, 2005).

•*Prion diseases*
The PrPsc is responsible for all the transmissible subacute spongiform encephalopathies, whether human or animal, such as: Creutzfeldt-Jacob disease (CJD), kuru, scrapie and Bovine spongiform encephalopathies (BSE). These prion diseases are transmissible in a natural or experimental manner but are not contagious strictly speaking in the usual sense (for example, by contact between a sick person and a healthy person).

The first disease which was demonstrated thereafter to be linked to the presence of a pathological prion was described in 1730. It was observed by some English breeders in flocks of sheep. It is known today as "scrapie". It was at the end of the 19th century that cerebral lesions, especially the presence of vacuoles, were depicted as being caused by the disease (C. Benoit). In 1936, two French veterinarians, J. Cuillé and P. L. Chenu, succeeded in transmitting the disease to animals by inoculation (they published their work in the

Comptes rendus de l'Académie des sciences, Proceedings of the Academy of Sciences). This research was taken up by an English veterinarian, Dick Chandler, who was successful in triggering the disease in a mouse by inoculation (1961). In 1996 (see above), Alper and Latarjet observed that the scrapie agent was resistant to high doses of ultraviolet and suggested that it was not a virus. Finally, it was in 1982 that Stanley Prusiner purified the infectious agent which he called "prion".

In 1920, while in the service of the famous neuropathologist Alois Alzheimer (the discoverer of the sadly famous neurological illness), Hans Creutzfeldt and Alfons Jacob described the symptoms of the disease which bears their names. The brains of the patients are riddled with microscopic holes (spongiform appearance) and contain deposits having the form of "plaques", as in Alzheimer's disease. The so-called amyloid plaques are protein aggregates (prions, neural debris). Creutzfeldt-Jacob disease (CJD) results in dementia and its outcome is fatal.

We know today that it can exist in several forms:

a) The most widespread is the form called "sporadic". Between 1992 and 2001, its prevalence amounted to 540 deaths in France. Its effect is most prominent after 64 years. Its cause is unknown. It could result from a spontaneous mutation of the PRNP gene coding for the PrP protein, or from a "spontaneous" conversion of the healthy form into a pathological one, etc. It manifests itself by early dementia which progresses rapidly, problems with equilibrium, visual anomalies and, histologically, by some cerebral spongiosis, a glial inflammation and some neuronal loss, with, however, an absence of PrPsc in the biopsy of the tonsils.

b) Much rarer are the "genetic" forms, linked to the mutations of the PRNP gene. These can affect the codon 200 (familial CJD) and are hereditarily transmissible; they can also affect this gene at codon 129 or 178, causing what is known as "fatal familial insomnia", which suggests that the "normal" protein would play some role in the waking-sleeping equilibrium. (Cases of sporadic fatal insomnia (non-heritable) have also been described.) Finally, the sporadic modification of codon 102 leads to Gerstmann-Straüssler-Scheinker syndrome.

c) Forms called "iatrogenic" are also known, those which, above all, have caused the greatest commotion in the world. They have led to safety measures of considerable magnitude (large scale slaughter of suspect herds of flocks, a ban on using animal meal to feed livestock, confinement measures at national or international level, embargoes on certain exports).

These measures and regulations have given rise to the creation of new "agencies"; for example, in France, the French agency for Sanitary Safety (Agence de Securite Sanitaire).

One particularly serious manifestation of contamination, having caused 79 deaths in France between 1991 and 2001, followed the administration of the growth hormone for the treatment of small size. Initially, the hormone was, in effect, extracted from the pituitary glands of human corpses, some of whom were contaminated. Some transplants of the cornea or dura mater have also created cases.

Kuru, a neurological disease (see below), can be placed in the category of the iatrogenic forms of CJD. This singular disease was described for the first time in 1957, by an Australian epidemiologist, Vincent Zigas, and an American doctor, Carleton Gajdusek. It is rife among the population of Papua New Guinea (2000 deaths were counted in 1950). Its very extensive study would lead to Gajdusek winning the Nobel prize in 1976 (Gajdusek had also previously established, on epidemiological bases, together with Françoise Catala, a link between the infectious agent of scrapie and that of Creutzfeldt-Jacob disease). The origin of kuru is the consumption of brains from human corpses (cannibalism).

However, fears were amplified when in 1996 a new CJD "variant" appeared, leading to the death of young people (aged less than 30) with histological signs distinct from those observed in the brains of people dying from the ordinary sporadic form. The lesions presented themselves as amyloid deposits surrounded by vacuoles ("florid" plaques). It has been established that the responsible agent was a prion responsible for bovine spongiform encephalopathy (BSE) and that, in all probability, patients suffering from the so-called "florid" form due to this variant CJD had contracted the illness after eating contaminated beef. Up to 2001, 103 cases had been observed in the United Kingdom, one in Ireland and three in France. We still do not know why the new variant affects younger subjects than the sporadic form responsible for CJD. The length of incubation, which in the latter case seems to be on average around 40 years, would be significantly shorter for the young people affected by this new CJD variant.

Individual genetic factors are probably also quite important[11] for the development and incubation period of CJD disease in general.

---

11. It was found that 70% of people with sporadic CJD are homozygous vis-à-vis the various coding forms of the triplet 129 in the PRNP gene. They belong to the type Met-Met (Met:

Much work has been undertaken to develop diagnostic tests that would detect signs of the fore-runners of prion encephalopathies, which would clearly be of great interest to asymptomatic people at risk, and also for veterinary examinations.

Currently, various studies are based on the hypothesis that specific distortions could be detected, for example, in blood samples by a systematic analysis of the plasmatic proteome or by transcriptome analysis. Different immunological procedures are equally being studied.

At present, when a clinician, confronted with a patient, is led to suspect the presence of CJD because of neurological signs or even some manifestation of dementia, he must call upon a range of examinations (tonsil biopsies, lumbar puncture, scrutiny for electro-encephalic periodic discharge characteristic of a sporadic CJD).

As far as the therapeutic approaches are concerned, they still remain quite limited in spite of the multiple studies achieved to date worldwide on a very large spectrum of candidate molecules, in an attempt either to block the synthesis of normal PrP, or to prevent its conversion into its spurious form. For now, none of these molecules can prevent the development of spongiosis or the fatal outcome of the disease, while some of them can lengthen the incubation period. Some trials have been carried out with "diapsone", an agent used in the treatment of leprosy. It has also been observed that an antifungal, Amphotericin B, would slow down, in a significant manner, the onset of the disease. Other substances such as congo red and penthosan polysulphate seem promising.

II.1.2. Genetic diseases – Gene therapy

Molecular genetics, under many different aspects, is in the phase of revolutionising medicine. If, as we will see later, it has not translated into as many therapeutic advances as marked as one could hope (apart from the enrichment of modern pharmacopeia due to biotechnologies inspired by genetics), it has achieved enormous progress regarding the mechanisms underlying

---

methionine), 15% are homozygous val-val (val: valine), the 129 triplet coding one or the other of these two amino acids. Finally, 15% are heterozygous. Among healthy people the values are 40%, 10% and 50%, respectively. Finally, all the people affected by the new variant are genotypically Met-Met (data supplied by INSERM in "Repère", Sept. 2006, p. 9).

many diseases, their nosology and the prediction of their onset. As the French biologists Jacques Ruffié and Jean Dausset argued, one can therefore speak about a genuine "predictive medicine"[12].

II.1.2.1 Historical aspects

It is mainly during the 20th century, as a result of progress in molecular genetics, then in genomics, that the widespread existence of a genetic "component" of many diseases has been highlighted.

Let us recall that the mutations involved can appear spontaneously, i.e. independently from the genetic patrimony inherited from the parents. We know, for example, that 3 to 4% of newborns are affected. Besides these spontaneous mutations, there is a large variety of mutations transmitted hereditarily.

The evidence for hereditary diseases and morphological or physiological syndromes is, however, quite ancient: in 1791, the "savant" and academic Maupertuis described the syndrome called "polydactyly". But it was the French physician Duchenne, a pupil of the great neurologist Charcot, who highlighted the first case of a muscular disease of early onset, which he called "myopathy" (later called Duchenne's disease, then Duchenne muscular dystrophy) (1868). Shortly after, C. Darwin (1875) described ectodermic dysplasia, and Huntington a certain neurological disease (chorea) which, from then on, bore his name. However, it is mainly the work of an English doctor working in St. Bartholomew's hospital in London, Archibald Garrod, who in 1901 made the first real diagnosis of a true "hereditary" disease, "Alcaptonuria", detectable by the darkening of the urine. This rather benign affection is in fact due to a mutation affecting the metabolic conversion of an amino acid, tyrosine. Garrod was able to establish that this flaw was hereditary and that its transmission obeyed Mendel's laws. A short time later, he described other hereditary affections of the metabolism such as porphyria, cystinuria, etc. (1909: In-born errors of metabolism). Phenylketonuria, which is still today systematically diagnosed because of its haematological incidences, was described in 1934. Decisive progress was accomplished in the understanding of the hereditary nature of blood diseases when it became possible to correlate the mutation at its origin with a precise molecular change. We owe this achievement to V. Ingram and H. Dintzis, two researchers from the laboratory of the great biologist Linus Pauling. In effect, they were able to establish that so-called "sickle cell anaemia" (thus named because the red blood cells of patients assumed a

---

12. The case of infectious diseases will be the subject of a specific chapter.

"sickle" morphology), a disease equally known under the name "drepanocy-tosis", is due to a mutation at the heart of the gene coding for the " β" chain of haemoglobin, that causes the replacement of a unique amino acid inside the β polypeptidic chain. L. Pauling used at that time to refer to this situa-tion as "molecular disease", but the term genetic disease became, from then on, common parlance. Today, we distinguish among genetic diseases, those which, like drepanocytosis, are due to a unique mutation in a well-determined single gene (these diseases are said to be monogenic), and those which are associated with the presence of mutations affecting several genes; the latter are called "polygenic" or even "poly-factorial". A whole series of syndromes such as asthma, obesity, rheumatoid arthritis, etc., which have a genetic component, belong to the latter category.

The characterisation of monogenic diseases has made immense progress during these last few decades. While during the period illustrated by work from V. Ingram and H. Dintzis, one could count them on the fingers of one hand, today more than 5,000 of them have been recorded. Their prevalence is generally low or very low, varying between 1/10,000 and 1/100,000 (or less), which has led to the concept of "rare diseases". This concept is not based solely on numerical observation. It also aims at attracting the attention of the public as well as the medical, pharmaceutical and political worlds to the necessity of not including these diseases – often serious, chronic and sometimes fatal – in the group of "forgotten" diseases (an "Institute of rare diseases", an Institute without walls, was created in France a few years ago). Their existence raises a genuine social ethics problem, for the same reason as physical or mental handicaps, since this low frequency of onset hardly encourages the phar-maceutical industry to make significant investments for the development of potential therapeutic treatments.

Moreover, if one considers the set of genetic diseases as a whole, one can estimate that in France, for example, they affect nearly three million people and at least 25 million in Europe!

II.1.2.2. The example of Duchenne muscular dystrophy (DMD) – A school case

The study of genetic diseases, as has already been emphasised, unde-rwent a rapid expansion at the beginning of the 1980s when Monacco and Kunkel succeeded, for the first time, in characterising the molecular target of a serious genetic disease, Duchenne's myopathy (also called Duchenne muscular dystrophy or DMD), a disease affecting the muscles called "skeletal" (or "voluntary" muscles, as opposed to smooth and cardiac muscles). In fact, the first work of the American group, quickly giving rise to the collective mobilisation of numerous teams, encouraged by various associations of patients and their families (in France, AFM: French Association against Myopathy), could establish that DMD is a disease linked to the X chromosome and that its target is a very large gene coding for a protein that was named "dystrophin". This protein, situated under the muscular membrane (or Sarcolemma) is of a very high molecular mass. It is made up of helical regions and of less structured domains. In the interior of the muscular fibre, it is linked with skeletal "actins" (a kind of contractile protein) and, at the surface, projected towards the exterior of the contractile fibre, it is connected to a complex network of polysaccharides and glycoproteins (or sarcoglycans).

The DMD gene can be the site of point mutations (often giving rise to changes in the reading frame, thus interrupting downstream the weaving of the polypeptidic chain during its synthesis). More often, DMD mutations are accompanied by deletions of varying lengths, some being able to alter the promoter(s) of the gene. Therefore, according to the nature of the modified sites, the lengths of the dystrophin polypeptide segments that will be synthesised will vary. When some of these segments are of sufficient size to give the muscular fibre a mechanical resistance enabling it to ensure, during some periods, the contraction cycles to which it is normally submitted, it will result in an attenuated form of disease called "Becker's" myopathy. On the other hand, when the synthesised dystrophin segments are not sufficiently long, the sarcolemmic membranes will soon present numerous rupture points, following an increased penetration of certain ions and a more or less pronounced proteolysis, often accompanied by an extended multiplication of fibroblasts. Other mutations can destabilise the structure in which the dystrophin participates by affecting the extra-membranous glycoproteins. They give place to other forms of myopathies (for example: sarcoglycanopathies).

In short, the genetic study of Duchenne's myopathy has strikingly illustrated the extent to which molecular biology (and the technological support involved in its development) is able to clarify the global phenotype of a genetic disease and consequently, to define its nosology better[13].

This situation is conveyed today by an unprecedented enrichment of the nosological framework of diseases of a hereditary type. Thus, before the first work of Kunkel and Monacco, already quoted, only very few myopathies had been described. Today, more than one hundred types of neuromuscular diseases (including mitochondrial myopathies) have been characterised. In a

---

13. Study of distinctive characters allowing the definition of diseases.

number of cases, the target genes have been cloned and their products identified. In others, they have been located with enough precision so as to exclude any synonymy[14].

II.1.2.3. Neurological affections

Today, the occurrence of human genetic diseases has been described for the majority of tissues or organs. For example, it is the case for numerous neurological affections concerning either the central nervous system (e.g. Huntington's; X-linked mental retardation; Parkinson's; Alzheimer's, etc.) or its peripheral counterpart (e.g. lateral amyotrophic sclerosis, spinal amyotrophies, Friedrich's ataxia, etc.). The affected gene in Huntington's disease codes for a "Huntingtin", a protein whose role is still unknown. The disease is manifested by a serious incoordination of movement (chorea) and generally only appears towards the age of 40. From this arises a serious ethical problem linked to the opportunity of informing the affected patients and their families.

Neurological diseases often comprise a rather complex genetic component. They may be due to, or associated with the presence of mutations in genes of a different nature, present on different chromosomes. Such is the case, for example, of Alzheimer's disease, affecting a significant portion of the population, usually after the age of 60. In most cases, the disease is consecutive both to mutations altering specific cell components (structural genes) and/or some susceptibility genes. Post-mortem examinations have revealed three general types of lesions in the brain (within the hippocampus, the tonsils, the cortex, or even the blood vessels). One often observes the accumulation outside the neurons, but in the close vicinity of so-called "senile plaques" or amyloid deposits (due to the aggregation of a β-amyloid protein containing 40-42 residues, the $A\beta_{40-42}$). These deposits are consecutive to abnormal cleavage of a naturally occurring precursor, present in the membrane, the component βAPP (β-amyloid precursor protein). The anomalous cleavage can be due to mutations altering the βAPP sequence, or in the proteolytic enzymes, the β- or α-secretases. However, other mutations were described in 1995 that are also responsible for family Alzheimer's transmission. They affect genes situated on other chromosomes, which encode

---

14. Some of these neuro-muscular diseases can present a complex phenotype. Such is the case of Friedrich's ataxia, which manifests itself by neurological attacks, a cardiomyopathy and sometimes diabetes. The target gene at the origin of the disease was identified in 1996. The mutations which affect it lead to a deficit of proteins, causing an abnormal accumulation of iron in the mitochondria. This anomaly, in its turn, leads to the production of "free radicals" which have toxic effects on various tissues, in which they can provoke degeneration or death (A. Munnich).

two distinct membrane components of cerebral neurons, the presenilins I and II. Some of these mutations lead to an increased production of $A\beta_{40\text{-}42}$. Other Alzheimer's-associated lesions have also been described, that cause the formation of odd structures inside the neuronal bodies; these are called "fibrillary tangles" or "paired helical filaments" (PHF).

Other neurological diseases (but also some non-neurological ones, like myotonia) reveal a very remarkable genetic mechanism based on the existence, within the genome, of short repetitive sequences. These diseases are said to result from "triplet expansion". In the case of the "X-linked mental retardation" syndrome, but also in other alterations in cerebral development, the super-abundance of short, repeated motifs is frequently detected. These include, for example, codon CGG (a mental syndrome called "Fragile X", J.L. Mandel), codon GAA (spinocerebellar ataxias), codon GAA (Friedrich's ataxia) or even codon CAG (myotonic dystrophy, spinocerebellar ataxias, Huntington's, etc.). In the normal state, these codons form iterative chains of from 20 to 30 motifs juxtaposed "in tandem" in the chromosomal DNA. However, it could happen that in certain families, the degree of iteration increases at each generation, a situation that might cause pernicious changes in the activity of essential genes situated in the close vicinity.

II.1.2.4. Susceptibility Genes – Polymorphisms and diseases – HLA genes

As previously reported, the mutations are not automatically respon-sible for the outbreak of a disease. They might not manifest themselves by any pathological phenotype, or at most they might cause a physiological non-acute syndrome. The likelihood of the appearance of the disease which could result will of course depend on the nature of the gene and its role in the physiological economy of the individual, and also on the nature of the change appearing in the "sequence" of the target gene. The intensity of the pathological manifesta-tion might also be conditioned by the environment and by the overall genetic terrain of the individual concerned. Thus, the same mutation appearing in two separate individuals and concerning the same gene might in one case lead to a serious illness, and in the other only cause a light syndrome or even no manifestation at all.

The "genetic terrain" is by definition complex. It can be associated with epigenetic characteristics (for example, the degree of methylation in certain parts of genomic DNA), but also with the status of susceptibility (or predispo-sition) genes and with the types and extents of genetic polymorphisms.

The existence of predisposition genes has been particularly well documented in the research devoted to the occurrence of cancers (see above) and
we have also seen examples while discussing polygenic diseases. There is a
particular family of predisposition genes which has been the object of intensive study in so far as the system to which these genes belong, the HLA system,
intervenes in a predominant way in the compatibility of organ transplants and
in the immunological recognition of self and non-self, but also because it is
linked to the degree of susceptibility and resistance towards various diseases.

• *HLA system and predisposition to diseases*
    The HLA system consists of at least half a dozen genes which are mostly
juxtaposed on chromosome 6 in humans. Among the principal genes such as
A,B,C,DR,DQ and OH, some are situated in the vicinity of the centromere,
while others are close to the telomere. Thus, in man, the pre-centromeric HLA
genes are said to be of class II, those of class I being located close to the telomere. However, there are a large number of other HLA genes occupying other
positions. The products of these genes are glycoproteins present at the surface
of the majority of the somatic cells. They act as some kinds of specific signals
in the presentation of cellular antigens to the receptors of the T lymphocyte
cells, some of which are cytotoxic.

    Each of these genes can comprise many variations in its sequence
(variants called "alleles"). These can be very numerous (e.g. several hundreds
for gene B) and since each of us receives a series of these "variants" from our
parents, the number of "combinations" of HLA alleles can reach extremely
high figures. According to Professor J. Dausset (who received the Nobel Prize
for Physiology or Medicine for his work on the HLA system (1980)), these
allelic combinations constitute the true "hallmark of genetic individuality of
each of us".
    The function of the HLA genes is to distinguish between what immuno-
logists call the **self** and the **non-self** in giving an organism the possibility of
immunologically rejecting what is foreign to it. As reported above, some variations (some alleles) of the HLA genes are linked with increased risk of diseases to
variable extents. Knowing such HLA allelic variants therefore permits, in principle, the establishment for each individual of the risk of developing such and
such a defined pathology. For example, an individual carrying the variant B_{27} of
the HLA-B gene is confronted with 88 times more risk of contracting ankylosing
spondylarthritis (sclerosis of the vertebral column). The presence of
the variant HLA-DQB_1 is linked with a probability 50 times higher of developing
retinopathy (birdshot type) and narcolepsy (tendency to drowsiness). Other
HLA alleles are known whose presence is associated with the development of
auto-immune diseases such as type 1 diabetes in childhood, myasthenia gravis

(a disease of the voluntary muscles following the pathological formation of anti-bodies directed against the muscle acetylcholine receptor), or even multiple sclerosis, rheumatoid polyarthritis, etc. Conversely, allelic variants which give a natural resistance to certain affections are known. Among these is HLA-DR$_2$, which protects the carrier individual against infantile diabetes or even HLA-B$_{53}$, which gives resistance to malaria, etc.

• *Susceptibility genes and SNP-type polymorphisms*
 Apart from the genetic variants of the HLA system, we know today a large spectrum of predisposition or natural resistance genes to various pathologies.
 Several of them are, for example, related to the probability of developing certain cancers. Thus, the APC gene (present on chromosome 5q21) is in part responsible, under certain allelic forms, for the onset of colon cancer with polyps (1% of familial forms). Some mutations in the BCRA1 gene (chromo-some 17q) are associated with close to 40% of familial forms of breast cancer in early development, often accompanied by ovarian cancer. A second suscep-tibility gene, BCRA2, on chromosome 13, accounts for other types of familial breast cancer. These are only some examples among many. The knowledge of predisposition genes is responsible for the development of a new form of medicine called "**predictive medicine**", medicine with two faces since, on the one side it should allow the early introduction of appropriate treatments in the life of people at risk, while on the other side, if the genetic informa-tion of an individual becomes systematised, even obligatory, it could lead to deviances of any kind, and it should at least be strictly controlled at the legis-lative level in every case. Finally, note that next to the susceptibility genes of a disease which can be the centre of mutations, giving rise to an increased probability of the manifestation of the disease, there are, spread throughout the genome, hundreds of thousands of limited changes or "<u>variations</u>" (so called), which may concern only a single DNA base. At the level of a unique nucleotide, these polymorphisms or SNPs, are, as we have already pointed out, associated, statistically speaking, either with an increased sensitivity to certain pathologies or to drug incompatibilities. We know that the location and the nature of these SNPs might differ according to the ethnic group and sometimes this has encouraged comparative research relevant to the genetic make-up of a population. In other cases, these polymorphisms can help to establish, within a given population, the effectiveness of certain therapeuti-cals. This <u>pharmacogenomics</u>, which is still in its infancy, could then augur a "<u>pharmacotherapy à la carte</u>" (the tracking in a card of SNP variants known to be responsible for drug incompatibility could therefore condition a target prescription).

II.1.2.5. Gene therapy – The gene as a drug and gene surgery

Since the beginning of the 1990s biologists (such as Anderson and M. Blaese) have tried to correct genetic diseases or to lessen their effects by using gene therapy, first among laboratory animals and then in humans. The idea of this intervention was simple: to introduce into the genome of a sick animal or a patient carrier of a mutation at the origin of a determined genetic disease (e.g. Duchenne's myopathy, mucoviscidosis, immune disease or specific cancers) an appropriate gene capable of compensating for the deficient function caused by the mutation and thus alleviating a physiological distortion or a developmental problem. The introduction of a "compensatory" or "correcting" gene is achieved by using a molecular vehicle, called a **vector**, to which this gene is artificially linked and one expects from this intervention that the "gene of interest" will be incorporated into the DNA of the patient to correct the phenotypic consequences of the mutation. In general, two methods of intervention have been used: in one of them, the cells from the patient (most often of medullar origin) are cultivated *in vitro*, treated by the recombinant "vectorised gene" then re-injected into the donor. In other cases, one uses the systemic way, that is to say, the treatment is carried out directly *in vivo*. The vectors are often certain viruses offering the advantage of being able to be easily integrated into the genome of the receiver. These viral vectors are previously "disarmed", i.e. viruses which have been stripped of their powers of replication. In a good number of cases, people have had recourse to retroviruses (for interventions *ex vivo*), to adenoviruses, or to the AAV (associated adenovirus) for direct *in vivo* therapy.

The number of trials carried out on animals – most often mice strains affected by diseases of well-determined genetic origin – is considerable. These tests have often been accompanied by spectacular results (e.g. the curing of mice affected by sickle cell anaemia).

However, gene therapy trials in humans have, apart from one remarkable exception, encountered serious difficulties. These could come from several causes: insufficient expression of the "transgene" (i.e. of the gene introduced artificially for therapeutic purposes), limitation in the capacity to produce the appropriate vectors, the incapacity of the transgene to become integrated properly in the genome of the host, production of cytotoxic lymphocytes or antibodies directed against certain chemical motifs of the newly expressed protein, etc.

• *The work of A. Fischer and M. Cavazzana Calvo*
However, we know that a gene therapy protocol can be applied with success in the human species, thanks to work carried out in 2000 by the

group of Alain Fischer and Marina Cavazzana-Calvo at the Necker hospital. These researchers have succeeded in curing young children suffering from a very severe disease leading to the incapability of the bone marrow to produce the lymphocytes T and NK (natural killers). When the sick children are not treated, it usually results in a serious lack of resistance to all sorts of infections; therefore, in order to ensure their survival, they have to be kept isolated in a confined atmosphere. The mutation responsible affects the synthesis of the receptor to cytokines (indispensable factors in the maturation of the lymphocytes). A. Fischer and M. Cavazzano-Calvo have proceeded to transfer a normal allelic (non-mutated) gene by means of a retroviral vector into the hematopoietic progenitor cells, sampled from these young patients, putting these transformed cells in a culture medium to make them multiply, and finally to inject them back into the sick children. This protocol has been repeated on a certain number of young patients, often with practically total and apparently lasting success, allowing the confinement to be abandoned. However, in two cases (in 2002 and 2003) the treated children died from leukaemia following the spurious integration of the compensator gene into the vicinity of a "proto-oncogene". From now on, it seems that the teams have succeeded in mastering the methods for appropriate integration of the alien gene, which should allow intervention on a larger scale with success.

Numerous programmes of gene therapy have been followed at the international level. However, the gene therapy approach is still scattered with difficulties (previously reported). It also has its limits. For example, the introduction of a very large gene such as that of dystrophin is difficult. Also, in some diseases such as Steinert's neuromuscular disease, despite the introduction of a "healthy" gene, the mutated gene of the receiver would continue to produce substances that are toxic to the cell.

• *The strategy of exon skipping*
That is why some biologists have sought other methods corresponding to what is now described under the generic name of gene surgery. Let us quote, for example, the use of molecules in preventing premature elongation during the synthesis of an essential protein, due to the presence of a "stop" codon. One solution is based on the use of an antibiotic substance such as tetracycline which, in modifying the properties of the ribosome, forces it to bypass the "stop signal". Tests with a molecule manufactured by an American firm (PTC therapeutic) are also in hand in the treatment of some forms of mucoviscidosis and Duchenne's disease.

Another technical approach, particularly ingenious, consists of using the antisense RNA to prevent the reading of a mutated exon by the ribosome, a reading which would lead, depending on the mutation, either to the formation of a "truncated" protein, or the formation of a protein distorted in its sequence. Certain researchers have had the idea (Goyenvalle *et al.*, 2004) of making the cell of a patient manufacture, from an appropriate genetic construction, an antisense RNA whose sequence is <u>complementary to the transcript corresponding to the mutated exon.</u> The antisense RNA comes therefore to pair with the defective exon transcript, <u>which spares the splicing system from integrating it into the mature messenger RNA.</u> The corresponding protein, which the treated cell will synthesise, will indeed be shorter than that skipped transcript, but it might often remain functional. Very encouraging results have been obtained by this technique, of so-called "exon skipping", in mice or in dystrophic dogs, with the restoration of a functional dystrophin.

Finally, in certain cases (such as that of Steinert's disease as already quoted), the mutated messenger RNA accumulates in the nucleus of the cell and acts as a toxic element. Some Canadian researchers have had recourse to the RNA interference approach and have succeeded in destroying this noxious RNA messenger. To summarise, we can see that there are therefore very many tracks which are presently being explored to overcome the genetic defects responsible for very different diseases. It is to be hoped, through the examples highlighted above, that in a not too distant future improved diagnosis will enable the implementation of treatments or at least more appropriate therapeutic support, and that the list of diseases amenable to gene therapy will grow longer.

II.1.2.6. Children's diseases and congenital malformations

A particular aspect of the human cost of genetic diseases which the international organisations (the WHO, CDC – Centre for Disease control – in Atlanta, the "March of dimes" programme, the World Alliance of Organisations in Favour of the Prevention and Treatment of Genetic and Congenital Affections, the International Genetic Alliance of Organisations of Relatives and Patients, etc.) are beginning to be quite seriously and actively aware of, concerns children's diseases and congenital malformations (i.e. "birth defects").

As Doctor Ysbrand Portman in particular has emphasised, a recent report (MOD, 2006) shows that the prevalence of all genetic and congenital diseases at birth reaches 8.2% of full-term infants in countries of low income, against 4% in the industrialised countries. It can be established that each year around 8 million children - that is to say, 6% of all births in the world – are born with a serious defect of genetic origin either fully or partially proven. More specifically, hundreds of thousands are born with defects occurring after conception, following such diseases as rubella, syphilis or iodine deficiency. These birth defects can be lethal. Among surviving children, these defects may be accompanied by psychological, physical, auditory or visual problems. An estimated 3.3 million children under the age of 5 die from these physiological disorders each year, while 3.2 million children who survive remain disabled throughout their lives.

The malformations or the most frequent genetic diseases from which children may suffer from birth are of various kinds. They can be congenital cardiac defects (more than one million cases per year), malformations of the neural tube (more than 150,000 cases), disorders affecting the synthesis or properties of haemoglobin, such as thalassemia and sickle cell anaemia (more than 150,000 cases), Downs Syndrome (around 200,000) or glucose-6-phosphate dehydrogenase deficiency (between 150,000 and 200,000 cases).

Concerted efforts are beginning to emerge at the international level, but probably the most obvious measure lies in specialist training in medical genetics. Setting up "genetic medicine services", integrated into the network of public health, is an urgent matter. This implies a continuum in monitoring, as regards care, the conception, the mother's health, childbirth, the health of newborns and young children, taking into account genetic defects or diseases, not to mention early diagnosis, risk assessment and family counselling.

II.1.3. STEM CELLS AND CELL THERAPY (A HOPE IN THE FIELD OF DEGENERATIVE DISEASES)

II.1.3.1. Developmental biology considerations

If there is one favourite theme in biology, it is that of development. The understanding of the first stages of this extraordinary and fascinating process which is the formation and morphogenesis of the embryo and its transformation into an adult organism has long challenged the imagination of the early physiologists and embryologists. What mysterious plans for the organisation are present at this gradual metamorphosis? How does the execution of this plan, inscribed in the fertilised egg, lead to a living architecture, that is proper to the species? How are the limbs, organs and tissues differentiated? Once more, genetics has made tremendous progress in providing clear answers to many of these issues. The discovery in the second half of the 20th century of development genes and in particular that of the **homeotic** genes in the "drosophila" fruit fly, mouse and man, has revealed the striking developmental mechanism involving "morphological segments", foreshadowing what, in a way, is equivalent to the development of the organisational "pattern" of the body: trunk, limbs and head, as well as the position of the wings and antenna for insects. Apart from these organisational genes (kinds of "architect genes"), which define the plans for the overall architecture of the body (see F. Gros: "Secrets of the Gene"), it was important to bring to light the "cascades" activating those that govern the formation of specialised body parts. In this respect, one of the processes of development and tissue differentiation the best analysed so far in these last few years, has been that of the skeletal muscles (myogenesis). A decisive step has been the discovery by H. Weintraub of (determination) genes of the MyoD family, of their role during the formation of the first "myoblast precursors", a formation that is followed by that of "myotubes", contractile "muscle fibres" and organised musculature. Today, scientists have made an extremely precise inventory of the nature and role of the principal regulatory genes intervening in the differentiation of the embryonic "mesoderm" (somites), to form the contractile muscles of different parts of the body.

However, understanding the integrated functioning of the regulatory and structural genes, which participate in a development network, is not sufficient to explain the formation of the body in its entirety. The human body, for example, consists of between 10^{13} and 10^{14} (from 10,000 to 100,000 billion!) cells. The development does not boil down to the execution (however perfect it may be) of a complex programme of genetic expression in time and space. Very many movements, invaginations, protrusions, etc., accompany the morphogenesis of the body, parallel to the acquisition of physiological specificities unique to the tissues and organs. We also know (D. Duboule, Ameisen) that apoptosis (programmed cell death) is also indispensable to the "sculpture of the living". For example, without the triggering of apoptosis at a precise stage of foetal development, which eliminates interdigital tissues in man, the extremities of our limbs would be webbed like the birds'!

It is felt that in order to establish a developmental biology that is able to take into account all these complex phenomena, research cannot overlook general and comparative embryology. It is essential to monitor the fabulous destiny of the first cells formed soon after fertilisation of the ovocyte, to know the morphogenic factors and their "gradients" of distribution in the embryo, to analyse the formation of the first pre-differentiated territories, precursors of adult tissues, and the various paths of the first differentiated cells.

In this respect, modern embryology has shed precious light, by providing the possibility to carry out some marking of the precursor cells, distinguishable according to the morphology of their nuclei (in accordance with the technology introduced by Nicole Le Douarin). It enabled one to clarify the early stages of development of the nervous system from the neural plate and to clarify numerous ways of differentiation.

However, for a long time, embryologists and developmental biologists have been tempted to reproduce the stages of tissue differentiation by using *in vitro* cultivation of precursor cells. This has led to characterising, in the developing embryo, the very early cells thus programmed to differentiate, or those which have the potential to form all the tissues of the organism. Thus was born the interest in stem cells. Placed in certain culture conditions, these cells only self-reproduce. However, on the other hand, when the composition of the culture medium is adequate – that is when the necessary factors are present – they can differentiate *in vitro* into a variable number of specialised tissues. Therefore, one can distinguish among stem cells those endowed with "totipotent", "multipotent" or "unipotent" differentiating capacities, according to the aptitude they have for generating all the specialised tissues (in man, approximately 200 distinct types), only some of them or just a single one. One can distinguish two major classes of stem cells: **embryonic stem cells** (abbreviated to **ES cells)** and **adult stem cells.** As their names indicate, the first are derived from young dividing embryos, a few days old; the second exist, in a very small number, within differentiated adult tissues. The embryos, sources of ES cells, result from the divisions of a fertilised ovocyte and ES cells are sampled shortly after the young morula stage, a stage, called **blastocyst,** consisting of an inner cell mass coated with a layer of epithelioid cells, the trophectoderm. In usual development, the "inner cell mass" is at the origin of all the specialised tissues of the foetus, while cells of the external layer will form the future placenta. In the human species, the blastocyst is generally utilised towards the 5th day after *in vitro* fertilisation, before re-implantations *in utero*, in the case of a medically assisted procreation (MAP). It is also from the inner cell mass, and after it is propagated in a culture medium, that the mammalian embryonic stem cells are isolated. Before describing their characteristics, we will give several insights into adult stem cells.

II.1.3.2. Adult Stem Cells

• *Blood stem cells*

As we have said, a very small number of adult stem cells are generally present within specialised adult tissues from an organism. The most well known are the bone marrow stem cells. Our blood is in effect a very active "factory" whose constitutive elements are being constantly renewed. Today, one can distinguish long-term hematopoietic stem cells (so-called LT-HSC) which are responsible for this renewal during the course of our lives. The differentiation of LT-HSC into blood cells and lymphocytic cells follows successive stages: firstly there is formation of a second generation of "short-term", hematopoietic stem cells which are multipotent, following which, ST-HSC differentiate into non-self-renewing cells, or MPP, the latter being the immediate precursors of the lymphocytic lineage (immunity cells) and of the myeloid lineage (red blood cells and platelets). The genes involved in this cascade of events are quite well known, as are the corresponding growth factors such as the cytokines or the erythropoietin; equally known are the antigens present at the surface of various cellular intermediaries, which allows them to be "sorted" if necessary. Bone marrow grafting has been in use for quite some time to overcome different forms of anaemia resulting from genetic diseases or from aplasias frequently observed during use of anticancer drugs or after radiation, because hematopoietic precursor cells are capable of re-forming all the deficient blood elements. Unfortunately, the genuine hematopoietic stem, type LT-HSC, is very difficult, if not impossible, to cultivate. Cytokines, such as GM-CSF factor (granulocyte – macrophage – colony-stimulating factor) or interleukine-3 have been subject to many trials to maintain the precious cells in a proliferative state. More recently, people have had recourse to the "stem cell factor" and "thrombopoietin" with partial success.

Besides the cells which are responsible for hematopoiesis within the bone marrow, the mesenchymal part of the marrow is rich in precursors of the endothelial tissue, a major component (with the smooth muscles) of the blood vessels. Certain authors have postulated the existence of common precursors of the hematopoietic lineage and of the endothelial system of blood vessels (hemangioblasts).

Other authors have reported the identification, in the medullary tissue, of real pluripotent stem cells (Verfaillie) with properties similar to those of embryonic stem cells. Although the bone marrow constitutes the real reservoir of stem cells destined to become hematopoietic, attention has recently been focused on other sources. In fact, the presence of such stem cells has been described in the peripheral blood, the placenta and the umbilical cord. The clinical use of hematopoiesis precursors is fairly diversified, from the fight against leukaemia, lymphomas, blood genetic diseases in general, various auto-immune diseases or, as previously reported, the accompaniments of cancer chemotherapy.

• *Other types of adult stem cells*
    Another category of adult stem cell has been very much in focus and has given rise to remarkable medical applications. This concerns stem cells from the cutaneous epidermis, themselves responsible for the formation of keratinocytes convertible into epidermic tissue and hair follicles (Howard Green). The possibility of cultivating them *in vitro* on a large scale has proved to be of great use in the treatment of severe burns.

However, although the hematopoietic stem cells and those of the skin have been by far the best studied, nevertheless, for more than a decade much work has also been dedicated to other types of adult stem cells. In this regard, one should mention studies devoted to the skeletal muscles, the dental pulp, blood vessels, endothelial tissue and smooth muscles, as well as the intestinal tissue, the cornea, the retina, the liver, the pancreas, etc. A particular mention must be made here of the recent work dedicated to myogenic precursor cells (apparently very close neighbours of the stem cells) (Montarras *et al.*, 2007), which offer important perspectives of cell therapy for Duchenne type degenerative dystrophies (see above). (A special category of multipotent stem cells at the origin of endothelial cells and skeletal muscles was reported to exist within the aortic wall (G. Cossu)).

• *Neural Stem Cells*

However, one of the most remarkable and unexpected illustrations of the field concerns the discovery of **neural stem cells** capable of regenerating nerves and glial tissues. Their existence escaped identification for a long time and their discovery is relatively recent (B. Reynolds and S. Weiss, 1992; C. Lois et Buylha, 1993; Ron McKay, 1997; F. Gage, 2000). It has raised (and continues to raise) great enthusiasm among the neurobiologists and in medicine, in the scope of developing a cell therapy directed towards different neuro-degenerative diseases. A dogma was shattered, according to which the neurons of the central nervous system are not renewable. The regions of the brain in mammals, where this neurogenesis has been brought to the fore, are the sub-ventricular zones (lateral parts of the ventricles of the forebrain). The stem cells existing in this particular zone generate *in vitro* neuronal progenitors which migrate to reach the olfactory bulb at the level of which they convert into granular neurons. When cultivated *in vitro*, the stem cells of the sub-ventricular zone form rounded clusters of cells or "neurospheres" before the appearance of the three major types of neural cells typical of the central nervous system: neurons, astrocytes and oligodendrocytes.

Other neuron-forming sites have been described in the hippocampus, where cells of the granular zone (which neuro-anatomists call the **gyrus-dentatus)** reside. One can also include in the category of neural stem cells, cells of the "neural crest" (N. Le Douarin) which are responsible for sympathetic and para-sympathetic neurons as well as many non neuronal tissues. Finally (although we are not strictly concerned with adult stem cells here), biologists have also observed that human foetal cells from the mesencephalic region could be used to alleviate, at least temporarily, the neuronal degeneration which occurs in Parkinsonians and leads to the death of the doparminergic cells (M. Peschanski)[15].

---

15. However, it is interesting to report that in 1999, the team of Ron McKay had observed that mice ES cells could, in certain culture conditions *in vitro*, differentiate to form a large number of dopaminergic neurons.

Before leaving the chapter of adult stem cells and turning towards the perspectives (and debates) surrounding the potential use of human embryonic stem cells, it might be worthwhile to address a series of observations on an unexpected property of this category of cells: their so-called plasticity.

• *"Plasticity" of Adult Stem Cells*
      The first article mentioning this property was published in 1999 (C. R. Bjornson, 1999). There, it was related that, in certain conditions, the nerve cells could adopt a phenotype of hematopoietic cells! Many examples in confirmation of this "plasticity" would be advanced afterwards. Thus, bone marrow cells, considered by their authors as authentic hematopoietic stem cells, would have, apart from their classical fate, the capability to form skeletal muscles in vivo, as shown by appropriate labelling criteria (Ferrari *et al.*, G. Cossu *et al.*, Muligan, Grusson *et al.*). They equally would possess the aptitude to form myocardial cells (Jackson *et al.*), hepatic cells (Patterson *et al.*, Lagasse *et al.*) or even nerve cells (Woodbury). Conversely, in some way, the formation of blood cells in aplastic mice (mice whose bone marrow is no longer functional) from muscle stem cells has been reported. Brain cells can generate muscle, etc. These different observations have been interpreted as proving the existence, under certain conditions, of a real phenomenon of "transdifferentiation". Certain authors (H. Blau) have hypothesised that in animals, some stem cells would circulate through the bloodstream; these cells having the potential to differentiate into varying phenotypes according to the nature of the tissue or organ that they reach, and the differentiation into that tissue or organ being stimulated when the tissue is injured and thus releases appropriate tropism factors. A variant of this hypothesis is that each tissular niche would host different types of adult stem cells endowed with distinct differentiation potential. It would be the environment created by the adult tissue with which they come into contact, in the last resort, which would determine the phenotypic fate of these stem cells.

      Without calling into question the actual observations related to these "plasticity" phenomena, nevertheless, recent data have led to questioning their interpretation. In fact, some recent publications seem to invalidate the hypothesis of transdifferentiation; the fusion of stem cells, belonging to a specific tissular niche, with differentiating cells coming from other tissues, would be at the least a plausible explanation!

II.1.3.3 Embryonic Stem Cells

      Whatever the considerable interest of adult stem cells at the fundamental level, and in certain specific cases their medical interest (cells of hematopoieticlineage or keratinocytes, etc.) also, the majority of scientists and, with them numerous physicians, attribute to the embryonic stem cells a greater potential to face the challenges posed by degenerative diseases, even if the ethical problems attached to their use are raising problems on a different scale.

• *Historical aspects*

The use of embryonic stem cell lineage (ES type) has been the culmination of quite a long experimental process. Without expanding, we should remember that the first characterisation of these cells was preceded by much research, mostly after the last world war, on a particular category of multipotent cells coming from cancerous tumours developed in mice germinative cells. From these "teratocarcinomas", biologists succeeded in isolating kinds of cells behaving like stem cells. When they were introduced *ex vivo* into a developing mouse embryo, they led, after transfer *in utero*, to the formation of young mice with a mosaic fur, whose tissues came either from cells of the carcinoma, or from normal embryonic cells. Placed in *in vitro* conditions, the embryonic carcinoma cells were able to differentiate into several types of tissues (skeletal muscles, myocardial cells, adipocytes or nerve cells) (F. Jacob, B. Mintz). Yet, in spite of their interest, further use for therapeutical purposes was hampered by their tumorigenic properties.

This obstacle has been lifted by the pioneering work of Gaël Martin, in the United States, and Martin Evans, in Great Britain. For the first time, they were able to isolate from mouse embryos cells which, when put in a culture medium, kept their ability to differentiate into many somatic cells or tissues even after multiple divisions. For many years, these murine ES cells have permitted many investigations dealing with embryogenesis *in vitro*. (Shortly after, they would also lead to the development of the gene invalidation techniques (so-called "knock-out") (Capecchi M.)). However, when people attempted to extend the observation achieved in mice to other models, it was realised that only a few species lend themselves to the establishment of permanent cellular lineage from the embryos, and for several years, the isolation of such lineages was limited to a unique mouse strain (mouse 129).

• *Discovery of human embryonic stem cells and potential applications*

However, in 1995, J. Thomson at Winsconsin University succeeded in cultivating ES cells derived from embryos of the Rhesus monkey; and finally, in 1998, he and his colleagues showed for the first time that the human embryo could lend itself to the isolation of precious ES[16] cells. From then on, as Nicole

---

16. Practically simultaneously, another American researcher, John Gearhardt, demonstrated that some pluripotent human cells (known as EC) could be obtained from germinal tissue (destined to form gametes) taken from the gonads of aborted human foetuses.

Le Douarin writes (in Regenerative Cellular Therapy, Letter of the Academy of Science n° 20 (2006)), "an inexhaustible source of unique human cells to replace dead or inefficient cells through bioengineering was available for medicine". The possibility of maintaining culture lines of <u>human embryonic stem cells (hEC)</u> has raised enormous hope in the medical world and also within the public. Embryonic stem cells, because they are pluripotent, offer in principle an ideal solution to alleviate, by cell therapy, serious physiological deficiencies such as, for example, type I diabetes caused by the immunological destruction of the Langerhans islands of the pancreas, or degenerative diseases, especially of a neural nature (Alzheimer's, Parkinson's, etc.) or even spinal cord injuries, all diseases or serious dysfunctions for which medicine is not yet prepared or for which treatments are long and complex. (Regarding type I diabetes and the hopes cherished by medicine of proceeding to a cellular therapy, it must be emphasised that while it has not proved possible up to now to characterise the presence of adult stem cells in the pancreas, on the other hand, scientists have progressed in the "transformation" of embryonic stems cells into cells producing insulin.)

Some research, regarding the properties of stem cells, has also been carried out with the objective of offering new strategies in the fighting of cancers.

For the first time, in 1997, researchers from Toronto University identified <u>cancerous stem cells</u>, by transferring blood stem cells from patients with leukaemia into mice and by showing that these recipient animals also developed leukaemic syndromes. Cells related to stem cells have equally been isolated from breast or brain tumours. Compared to what is observed in stem cells of healthy tissue, tumour stem cells exist in very small numbers, but they replicate, easily giving rise to a multitude of daughter cells. However, contrary to normal stem cells, they are insensitive to the regulatory mechanisms leading to the stopping of their divisions! Classic chemotherapy is capable of killing the majority of the tumourous cells, but when a few cancerous stem cells survive this treatment, a new cancer can develop. Research, focused on the observable differences in the range of genetic expression between normal and cancerous stem cells could turn out to be of great importance in preventing this type of recurrence.

Other applications arising from the medical use of stem cells are conceivable, distinct from the opportunities offered by <u>cell therapies</u> or from the study of cancer cells.

For example, stem cells could revolutionise traditional "chemical medicine". In fact, since the embryonic stem cells can differentiate *in vitro* into a large spectrum of specialised tissues, this should offer some possibilities for testing the effects of numerous pharmaceutical agents on these tissues without calling for healthy volunteers. The technique of nuclear somatic transfer (see below) applied to stem cells could turn out to be invaluable in the aim of exploring the effects of new drugs on diseases of genetic origin. For example, it is indeed difficult to study the progression of Alzheimer's or Parkinson's disease in the brain tissue of living patients. However, by using the cells from an Alzheimer's patient to create lines of stem cells, after nuclear somatic transfer, it would be possible to follow the development of the disease *in vitro* and to test chemical agents capable of regenerating nerve cells.

• *Risks*

While they may not have toned down the hopes placed in cell therapy (which remain very much perennial), several difficulties – some biological, others ethical – have at least shown that complementary research is necessary and that an ethical social consensus is imperative before moving to the clinical stage.

Regarding the potential difficulties (or biomedical problems) to be solved, one of these is that the injection of specialised progenitor cells into a patient could be accompanied by a malignant drift, since a small percentage of human progenitor cells could still carry undifferentiated stem cells. This problem has been the object of many studies. Therefore, a cell therapy protocol should, by no means, involve injecting the stem cells themselves into a patient, but rather the first differentiated elements (precursors) which emanate from them.

Another difficulty could reside in the involuntary transmission of viral pathogens from animals to humans, a possibility linked to the fact that the stem cells are cultivated in a medium enriched with products of animal origin (such as growth factors or serums). Hence, there is the necessity of developing culture media that are entirely synthetic (for example, cultures on "matrigel"). However, what was feared above all was the risk of immune rejection of the injected cells. In fact, we have seen that this risk is increased when the transplanted cells and those of the receiver are not histocompatible (see chapter on the HLA antigens). Although this difficulty could be circumvented in a mouse (from whom a large number of histocompatible lines are derived), the situation is not as favourable in humans. This is the reason why an experimental protocol supposed to avoid this major difficulty has been conceived. It is based on the replacement of the genetic equipment (the nucleus) of the embryonic stem cells by that of the future receiver, in such a way as to ensure a perfect immune compatibility. This technique of nuclear transfer, with a therapeutic aim, is called somatic nuclear transfer (sometimes incorrectly designated as "therapeutic cloning").

• *Somatic nuclear transfer (therapeutic cloning) – Reproductive cloning in animals.*
The nucleus of a somatic cell taken from a tissue of the future receiver (for example, from the epidermal tissue) is transferred (by electroporation) to the cytoplasm of an ovocyte which has previously been enucleated. The ovocyte thus reconstituted, containing the somatic (diploid) nucleus, is cultivated *in vitro* up to the blastocyst stage. Then, the process is as for a normal blastocyst; the cells of the inner cell mass serve to establish the lines of pluripotent stem cells, save that their phenotype is from now on compatible at the immunological level with that of the patient.

This process is sometimes described under the name of "therapeutic cloning". Its only justification, if it were to be put in practice in humans (which is still not the case), would in effect be of a therapeutic nature. It would not be based on an intention of cloning any human being. In an animal, if an enucleated ovocyte receives the somatic nucleus of a donor, following which it is placed in the uterus of a female carrier, an offspring can be obtained which will show all the characteristics of the donor. This operation, called "reproductive cloning", was first practised successfully in sheep; it gave rise to the birth of the famous sheep, "Dolly", produced by the Scot Ian Wilmuth (1996). Other animal clones have been produced in cows with a fairly good yield by the INRA team directed by Jean-Paul Renard. However, this operation is itself quite delicate. Numerous animals issued from reproductive cloning often manifest serious anomalies and in most cases, only a small percentage of them reach the term of complete development.

Although no case of human reproductive cloning has thus far been reported, this operation being strictly prohibited at the international level, nuclear transfer, albeit conceived with therapeutic aims, has also been banned in different countries, including France. This is in spite of the fact that the blastocyst containing the somatic nucleus is not destined to be implanted and that one is only dealing here with an intermediary step in the obtaining of immunocompatible embryonic stem cells (the UK and South Korea have authorised the practice). It should be noted that up to now no nuclear transfer approach aiming at producing human embryonic stem cells has met with success (despite some publications that have been invalidated thereafter). Nevertheless, scientists think that should the technique become successful, it would offer among other possibilities that of studying *in vitro* the development and progression of some specific diseases, the stem cells thus obtained having the pathological genotype.

From a different angle, the success of reproductive cloning, as achieved in different animal species (J. P. Renard), clearly indicates that after its transfer in the cytoplasmic context of an ovocyte, the nucleus from a differentiated somatic cell (for example, a skin cell) could undergo total reprogramming, making it acquire the characteristic of totipotentiality. This "reprogramming" mechanism is very much the focus among developmental biologists (H. Blau). It is without doubt very complex and involves epigenetic modifications which are only just beginning to be clarified.

II.1.3.4. Ethical aspects of the use of embryonic stem cells

The potential therapeutic use of human embryonic stem cells has raised, and continues to raise, numerous ethical questions. It has provoked strong debates in the world and has led to rules and laws being implemented. It is outside the field of this present book to trace their history or to analyse their different aspects. We will only recall below the principal implications.

One of the major criticisms deals with the method of obtaining human ES cells, as the establishment of a culture of hES cells is accompanied by the destruction of the blastocysts, i.e. young embryos (actually human pre-embryos). Yet, if these latter were implanted in the uterus, as is the case, for instance, in medically assisted procreation (MAP), they could generate a complete human being. The argument put forward by biologists and physicians to counter this view is that the human blastocysts might well end up being destroyed in any case! In fact, as stipulated by the French Bioethical Law, frozen blastocysts (often called "surplus embryos") must be destroyed after a few years when they are no longer the object of a parental project (Law of 1994). However, this destruction of embryos, whether linked or not to the production of hES cells, is considered by some opponents as unacceptable, with the argument that the human embryo acquires the status of a human being from the very moment following fusion of the gametes. Such is especially the position of the Catholic Church and of part of public opinion. For other religions, this statute is only attributable to a stage of development situated well beyond that of the "blastocyst".

In a more recent version, the French Law of Bioethics (2004), while maintaining the ban on what it describes as "embryonic research" and especially, the production of hES cells from surplus embryos, has, however, allowed some exceptional situations. Certain laboratories are now given the right to use embryonic lines for their research as long as they are used for therapeutic purposes and they are already established lines imported from abroad[17].

Recently, alternatives to the techniques used in deriving hES cells from surplus human embryos have been proposed. For example, some biologists report having been successful in establishing stem cell lines from embryos taken at the "morula" stage, under conditions that do not impair the continuation of normal development. Similar attempts have focused on the blastocysts. Other authors have succeeded in isolating stem cells from human amniotic fluid. (Dario Fanza, *Nature Biotech.*, 2006). Although they do not show the same degree of pluripotence as blastocyst-derived embryonic stem cells, these cells can be easily cultivated in the laboratory; they differentiate into a wide range of tissues: muscular, bony, myocardial, neural, etc. Furthermore, their use would present less risk concerning carcinogenesis.

Finally, but most importantly, a major advance has been accomplished, that might perhaps offer a substitute for the nuclear transfer technique described above. Two biologists, Yamanaka (Kobe University) and Thomson (Wisconsin University), have succeeded in converting true somatic cells (for instance, skin fibroblasts) into pluripotent cells harbouring properties very similar to those of embryonic stem cells (high dividing capacity, and ability to differentiate *in vitro* into a large spectrum of tissues, etc.). This was achieved by transfecting fibroblastic cells by a set of genes known to be involved in signalling pathways and immortalisation. These so called induced pluripotent cells (or iPS) are very much the focus in many laboratories. Although there is still some way to go before an application to human cell therapy, this technique of direct experimental reprogramming has already proved very useful in the hands of developmental pathologists (for example, in the establishing of cell banks corresponding to defined pathological features).

II.1.4. AGEING – SENESCENCE AND CELL DEATH (APOPTOSIS) – CANCERS

II.1.4.1. Ageing – General Considerations

If there is one phenomenon with the deepest repercussions in the future of our species and of the socio-economic development of the world, it is that of ageing of human populations. Here, many processes are brought into play,

---

17. However, a recent French parliamentary report (2005) strongly recommends shifting from such a dispensary regime to a clear authorisation of research on embryonic stem cells (Fagniez P.L., 2006).

ranging from the most diverse physiological manifestations concerning cells and tissues of organisms, as well as genes, their mutations and their polymorphism, to the demographic evolution of the planet!

In fact, demographers remind us that around 1750, with the sanitary conditions, food, etc., which prevailed in Europe, life expectancy at birth was only... 27 years. Only 21% of the population reached the age of 60. Those who did achieve what at the time was considered to be a physiological performance, could hope to live another 12 years. It has been established that 200 years later (in 1950), 70% of people reached their 60th birthday, with an <u>additional</u> life expectancy of 15 years on average. It is this last figure which has progressed since 1950; the life expectancy for a 60-year-old man today exceeds 20 years; it is significantly higher, close to 23 years, for women. It will exceed 26 years for men in 2050, and 31 years for women.

As Henri Léridon, member of INED[18] and the French Academy of Sciences, points out: "It is therefore the rise in the <u>longevity</u>, an <u>individual</u> characteristic, which becomes responsible for the acceleration of the <u>ageing</u> of the population, a <u>collective</u> characteristic". It has now become standard to speculate on the evolution of the number of centenarians, even that of "upper centenarians" (110 years and more). As a result, one wonders about the limits of human life by imagining, more or less implicitly, any form of intervention that would halt the process of senescence or degeneration of the tissues and organs (e.g. by the use of stem cells?). As Henri Léridon evokes in an imaginative but realistic fashion, "Our societies are entering into a totally unknown era, with the cohabitation of three, four, even five generations. Are they ready for it?"

However, faced with this situation and these questions which characterise in a general way the countries of the North – at least the majority of them – one can only be concerned about the serious discrepancies seen in the different countries of the South, especially in sub-Saharan Africa, as well as in certain regions of Latin America or South Asia. There, the joint factors of epidemics and bad food conditions, but also genocides, slow down the increase in life expectancy at birth, or even make it decline. The influence of the "diseases of poverty" and more particularly of AIDS, is often mentioned in this respect, even though recent epidemiological and demographical studies show that <u>in certain countries</u>, we can see a reversal in the curve, leading to a fall in infant mortality which is apparently linked <u>to factors other</u> than AIDS[19].

---

18. INED: National Institute of Demographic Studies.

19. Miniforum 2006 of COPED (the Committee for the developing countries in the Academy of Sciences) dedicated to the increase in infantile mortality in developing countries.

More generally, in fact, apart from the consequences of AIDS, we must often blame the lack of hygiene, the shortage of drinking water, the near absence of health centres or lack of transport.

Moreover, we know that the devastation caused by certain diseases prevalent in the southern hemisphere, although wholly and annually responsible for multiple early deaths, do not invoke much opinion and raise little international intervention. We have already mentioned these under the name of "neglected tropical diseases".

To return to the heart of the problem of ageing, which is multifaceted and poses considerable challenges to tomorrow's society, we are led to make a wish that the "Millennium Goals", aimed (among many of these goals!) at reducing, for example, deaths of children less than 5 years old, can really be achieved (this only happened in part). We are confronted here with two complementary approaches:

a) to ask science to understand the mechanisms of the phenomenon of ageing better, so as to alleviate the physiological disadvantages inherent in it, by fighting especially against chronic multifactorial diseases;

b) to turn towards representatives of the socio-economic and political world, so that it either acts precisely to help research and medicine, or that it acts, irrespective of this direct long-term approach, to improve the quality of life of older people (especially those of the 4th age, who are often "dependent").

While being aware of the different angles of study of the problem and taking into account the particular character of the present book, we will focus principally on its scientific aspects:

1) Is there a genetic determinism of the <u>longevity</u> of living beings and of man in particular?

2) What do we know about <u>ageing</u> in terms of molecular mechanisms, cell death or the "programmed" nature of this phenomenon (apoptosis)?

3) Finally, taking into account the relationship between the increased probability of the onset of cancer with the advancement of age, we will discuss certain facts concerning the link between genetics and cancer.

II.1.4.2 Genetics and longevity

For a long time, opinion has prevailed that ageing (and also death) has hardly had "anything to do" with genetics. That seemed to be justified, taking into account the enormous importance, in these phenomena, of environmental factors (nutrition, physical exercise, stress, accidents of all kinds). Besides, the existence of considerable variations observed from one individual to another, within a given species, and even more from one species to another, is striking. Yet, far from minimising the obvious environmental component (conditions and influence of the quality of life on longevity), biology and medicine are beginning to highlight the very important role, albeit still insufficiently understood of genetic factors.

Before coming to the relationships between genes and longevity (and between genomes and the average life span of the species), let us recall that, following Brown Séquard (1817-1894), Voronov and many others after them, biologists and physicians have for a long time emphasised the role of endocrinological factors. This interest is far from being extinguished. Much work continues to be dedicated to them, especially regarding the role and action of DHEA[20] (E. E. Baulieu). However, it was only towards the end of the 1980s and even more at the start of the following decade, that different laboratories discovered among laboratory animals genetic mutations capable of affecting, sometimes in a pronounced manner, the individual longevity vis-à-vis the other members of the species.

One can, along with F. Schächter *et al.* (1993), consider in fact three categories of genes as being capable of intervening in longevity: 1) those whose effects are globally similar among different species because they are linked with essential metabolisms; 2) those who play a role in the maintenance of the cellular integrity and in the repair of injuries; 3) the genes of susceptibility or resistance to age-related diseases.

However, many other classifications have been proposed (see, for instance, Pete Medavar, Georges Williams, George Martin, etc.). That of George Martin draws attention to the fact that although certain alleles (state of the sequence of a given gene) can have beneficial effects, sooner or later during the course of development, the same alleles can be repressed for "good reasons" or activated for bad ones during the course of ageing, etc.

The general developmental conditions therefore play a key role; they can considerably "modulate" the action of this or the other longevity gene. Once again, one observes that a gene should not be taken separately, when attempting to understand its effects, but in the overall context of (the effects

---

20. DHEA (dehydroepiandrosterone) is a hormone naturally produced by the organism. This compound is a derivative of cholesterol: a steroid. It is principally synthesised by its sulphate form: "DHEAS" by the surrenal glands.

of) other genes, individual polymorphisms and of course, environmental factors.

The first mutation affecting the longevity of an animal organism - in this case, the worm *C. elegans* – was described in 1988 by Friedman and Joberson. It was named "age-1" and it was found that it is associated with a doubling of the average lifespan of this nematode. Other mutations acting on longevity were described between 1988 and 1995 (e.g., Kenyou *et al.*; Larsen *et al.*; J. Z. Morai *et al.*; Kimura *et al.*). These mutations lead to a lengthening in the lifespan of the nematode worm by a factor between 40% and 100%. They also cause resistance to environmental stresses (thermal effects, ultra-violet and free radicals). It has been established, in fact, that the product of the age-1 gene, namely an enzyme, acting as a key element to cell signalling (in this case phosphatidyl-inositol-3-kinase), works by activating another gene, daf-2, which codes for an insulin receptor. Perhaps this result is reminiscent of an ancient physiological observation, according to which a lower consumption of sugar in the ordinary diet has a beneficial effect on longevity!

Other mutations observed in the nematode have effects on its life span. This is the case of the CLK mutation, a homologue of which has been observed in yeast, namely CAT5/COQ7. The mutations in question affect the transcriptional regulation of genes involved in energetic metabolism and, in some double-mutant daf-2-CLK, a considerable increase (up to 5 times) has been described in the average lifespan.

In a general manner (according to a recent summary by Doctor Richard A. Miller, of Michigan University) studies dealing with mutations related to the longevity of the nematode worm, and more recently of the drosophila, have shown that "monogenic" changes can increase the average lifespan by 100% or more. What is striking, is that the majority of these genes are involved in energy metabolism, and even more, in the resistance to cellular stress caused by heat, UV or free radicals (see the case of age-1), suggesting that the resistance to stress could represent a common characteristic of longevity.

Other work has been achieved in the mouse. Five monogenic mutations associated with an increase in the average lifespan by 30% to 40% have been described. In 4 out of 5 cases, it was found that these led to a significant reduction in the levels of the growth hormone and of its mediator, the factor IGF-1.

All the mutations so described correspond to the first category of longevity genes according to Schächter's classification (see above). Other types of mutations have been observed which could be placed in the second category as being linked to the maintenance of cellular properties. One of these, for example, is the gene whose mutation causes the Hutchinson-Gilford syndrome, better known as "progeria". We know that this syndrome, observable in humans, is expressed by the appearance of a whole series of symptoms typical of premature ageing. The gene in question is very close to the Rec-Q gene coding for a DNA-helicase and we know that its loss of function leads to a slowing down in the DNA replication, plus a general accumulation of mutations and an accelerated shortening of the telomeres.

The third category of genes able to play a part in longevity, according to Schächter's classification, consists of various diseases' susceptibility genes, as well as genetic polymorphisms. We have seen different examples of them previously, relating to HLA mutations. We will have the occasion to discuss them later.

While, as we see, genetics already brings certain interesting insights into the possible mechanisms by which genes can control ageing, for example by acting on the degree of resistance to physical or physical-chemical stress (see above), on the one hand, the hard facts remain insufficient to launch a general theory on the genetic determinism of longevity, and on the other hand we still do not know the nature of the genetic clock which makes ageing a biological phenomenon whose onset varies to a considerable extent from one species to another. The comparison between humans and mice is particularly striking in this respect; it concerns two mammals whose genomes are very similar: the total number of genes does not differ much and their sequences are equally close. In humans, at least in countries enjoying elevated norms of public health, the average life span is around 75 years, while in laboratory mice, according to recent studies, it is only 761 days (a little more than 2 years). This gives a difference of a factor of 36. As S. Edelstein (2005) points out, "very probably significant differences in a certain number of crucial genes are responsible for this factor of 36, but to be able to identify them is a huge problem".

• *Relationships between genomics and longevity in the human species*
A possible approach is to study the situation in individuals whose lifespan is clearly superior to the average in the considered species. Research has therefore been carried out with a view of identifying possible genetic features among centenarians, or even supercentenarians. The Study Centre of Human Polymorphism (CEPH), created about 30 years ago by Professor Jean Dausset in France, has led various studies on this subject. Others have been carried out: at the Deaconess Medical Centre (Beth Israël Hospital) and at the Hospital for Sick Children in Boston, as well as at Harvard Medical School, at

the Whitehead Institute and at Rutgens University, etc. In these various places interest was focused on the genomes derived from siblings of individuals with exceptional longevity (98 years or more).

For example, A. A. Puca *et al.* (2001) observed, by genotyping, a significant liaison between this exceptional longevity and a locus of chromosome 4 (D4S-1564). The nature of the gene(s) involved has not yet been determined. Other research, with a similar scope, was initiated in 1990. Thus, the IPSEN foundation has undertaken an important research programme, called "Chronos" involving 800 centenarians. It has shown a higher frequency of two particular alleles belonging to the gene coding for the apolipoprotein E and to the ACE gene encoding the angiotensin-converting enzyme. The APOE protein is a constituent of the lipoproteins playing a role in the transport of cholesterol towards all tissues.

Now, we note that in the centenarians, the APOE gene is predominant under its allelic form E_2 (an allele whose effect is to lower the serum level of cholesterol, contrary to the allele E4 which has the opposite effect and represents, for example, the most frequent form in people suffering from Alzheimer's disease).

As regards the ACE gene, it controls the synthesis of the conversion enzyme of angiotensin and bradykinin, two substances having a strong hypertensive action. It is the homozygous genotype D/D which prevails in centenarians, somewhat paradoxically because the allelic form "D" of the ACE gene is generally associated with a higher risk of coronary thrombosis!

We will conclude this sub-chapter dedicated to longevity genes by referring to some relevant perspectives. As certain physiologists have pointed out, "demographic projections have shown that a (hypothetical) intervention which could slow down the ageing of people, to the same extent as calorie restrictions slow down ageing in rodents, would have around twice the effect, in terms of the lifespan of a human in good health, than the complete treatment of all cancers, cardiac diseases, diabetes or strokes, considered globally".

• *The causes of physiological ageing*
    Even though genetic factors must be taken into account in what determines our lifespan, together with environmental factors, the mechanisms underlying the inexorable senescence of individuals within a given species, as well as their death, call for biological explanations. What, in some way, is the limiting biological factor? Certainly, there is the weakening of many physiological functions (cerebral, cardiac, respiratory, etc.) but saying so

does not evade the problem! Why, if purely accidental cancers or extreme environmental conditions are put to one side, does one witness different forms of physiological ageing? Should we not rather search for the cause in the "privacy" of the cells? And, to further deepen the research into the basic causes of biological ageing (in the limits of life expectancy proper to the species), should we not equally look into some kind of "wear" of the macro-molecules which they shelter?

• *Molecular ageing – effects of free radicals*
    In this respect, we know, since the early work of D. Hartman (1956) and the more recent study (1998) of K. Beckman and B. Ames, that the accumulation of free radicals can lead to different types of damage to a living organism during the course of its development.

This is the case of radicals of highly reactive oxygen, such as the hydroxyl radicals (OH) and the non-ionising radiations (which increase the content of this type of radical). Certain scientific arguments do indeed indicate that the effects of these free radicals intervene in longevity, in a negative way. Some of these arguments have already been evoked (as in the case of the mutation age-1 which reduces the sensitivity to radiation in *C. elegans* while increasing its lifespan). Other studies have focused on the drosophila. Thus, drosophilae, having received by transgenesis a gene coding for an enzyme destroying free radicals, such as superoxide dismutase or catalase, have a longer lifespan than that of non-treated drosophilae. In the same vein, in the age-1 mutants of the nematode, higher levels of enzymes with antioxidant properties have been observed.

The action of highly reactive free radicals on nucleic acids – DNA in particular – is conveyed by chemical changes capable of being eliminated by the repair enzymes. However, these "reparases" can be affected in their turn, as is observed in some cancers.

As regards the proteins, they can equally sustain multiple damage sites from the free radicals, manifested by biochemical changes such as glycations or oxidations. These damaged proteins are generally "vacuolised", then destroyed by an astonishing scavenging system called the "proteasome", which was discovered quite recently. However, according to work by Friguet (2000) as well as I. Petropoulos, the proteasome activity would decline during the course of ageing, which would lead to an accumulation of damaged proteins, especially in the keratinocytes (the skin precursor cells).

II.1.4.3. Cellular senescence

It was long believed that cells capable of being cultivated *in vitro* enjoyed immortality as long as the culture medium was renewed. Today we know that apart from the cells coming from what are called "established lines" (cells whose caryotype, i.e. the number and form of the chromosomes, is not entirely normal), and apart from cancer cells, the majority of cultivated cells have a limited capability for division, regardless of factors related to the medium.

One of the "cellular clocks" which takes part in this limitation lies in the gradual shortening of the telomeres, which is produced at each division. The telomeres, present, as their name indicates, at the end of the chromosomes, are formed by the repetition of a short sequence of DNA, generally TTAGGG. These sequences are organised as tandem repeats, i.e. juxtaposed. Yet, it has been established that at each new cell division, the telomeres become shortened (approximately 100 base pairs at least in each division). When the shortening of their sequence reaches several thousand bases, the cell ceases to divide and becomes "senescent". However, there exists an enzyme, the telomerase, which is capable of reconstituting these telomeres, at least partially, by lengthening them after they have lost some of their motifs. However, these telomerases are only present in developing embryos, and in the germinative cells, but are absent from other adult cells; they are equally active in the embryonic stem cells and in cancerous cells, which is not unrelated to the capability of these cells to divide *in vitro* in a nearly unlimited way.

Telomere shortening therefore plays a significant anti-tumoural role and the inhibition of the activity of the telomerase can constitute an anti-cancer strategy.

II.1.4.4. Apoptosis – programmed cell death

As we have just seen, our cells (apart from the case of cancerous or embryonic cells) only have a limited potential to divide and can become senescent when this potential is reduced or nil; the cells of old people have a lower dividing potential than those of adolescents (see review of G. Auboire, in *Biologie et Géologie*, APBG, n°3, p. 485 (2003)).

However, there is another cellular phenomenon, apparently parado-xical, whose importance was only revealed a few years ago and which plays a considerable role in the "normal" development of living beings or, on the contrary, in the appearance of cancer.

This phenomenon is known under the name of <u>apoptosis</u>, sometimes called "programmed cell death or cell suicide".

As biologist Jean-Claude Ameisen writes (in Annales de l'Institut Pasteur, Actualités, Apoptose en pathologie humaine, 11 n°4 (2000)), "Today we know that all our cells possess, during the lifetime of their existence, the power to self-destruct in a few hours and their survival depends, day after day, on their capability of "reading", in the environment of our body, some signals emitted by other cells, which they alone allow them to suppress the outbreak of their self-destruction. It is from the information contained in our genes that our cells produce the <u>executors</u> capable of precipitating their end and the <u>protectors</u> capable of neutralising these executors, and in a contra-intuitive way, a positive event – life – seems to proceed from the continual negation of a negative event, self-destruction".

It was only at the beginning of the 1970s that the real significance of apoptosis began to be perceived (J. J. Kerr *et al.*, 1972). Nevertheless, the phenomenon of cell death had been described for a long time. In 1842, J. Vogt, who worked on the metamorphosis of the tadpole, did its first description. Then it was the turn of A. Weissman to report on it during insect metamorphosis. In 1903, M. Ernst and A. Gluckmann established that cell death was a general phenomenon during the embryogenesis of vertebrates. J. J. Kerr distinguished two distinct types of cell death, at the morphological level: one type, called necrosis, an accidental and rapid death (i.e. unscheduled) which leads to the explosive disintegration of the cell, and the other, designated as <u>apoptosis</u>. The latter is expressed by some cellular condensation, the formation of invaginations, followed by implosion and phagocytosis. However, the cytological signs are numerous. To quote only the most obvious (E. Jacotot *et al.*, 2000), one can note, at the morphological level, a reduction in the cytoplasm, a pycnosis of chromatin, followed by a fragmentation of the cytoplasm and of the nucleus into apoptotic bodies. At the biochemical level also, apoptosis involves the following cascade: a fall in the mitochondrial transmembrane potential, an exposure of the phosphatidyl serine molecules to the outside of the cell, an activation of the proteases called "capases" (aspartate cysteine proteases), a splitting of the DNA into high-molecular-weight fragments or into oligonucleotides, etc. At this stage, the cell, thus greatly damaged, is finally phagocytised by the macrophages. This cascade of events, whose real start is the shutdown of mitochondria, which cuts, somehow, the energy fueling of the cell, is <u>strictly programmed</u>. In fact, it only intervenes in specific circumstances. These correspond either to <u>an act of defence by the whole organism</u>, which by a strategy of so-called "burnt earth" eliminates or tries to eliminate the cells which are victims of viral attacks, or which have become cancerous (we could in this case equate defence apoptosis to a form of cell suicide), or to a process of normal morphological or physiological develo-pment. Here, according to the formulation of J. Cl. Ameisen, apoptosis plays an indispensable role in the sculpture of the living.

Numerous illustrations pertaining to the role of apoptosis in biological development follow on from observations made on a large spectrum of species ranging from the nematode worm to man.

For example, the adult form of the worm *Caenorhabditis elegans* consists of exactly 737 cells (!) while the larval form has approximately 924. Therefore, apoptosis has caused some cellular loss during the formation of the principal adult tissues.

The metamorphosis of insects and birds is also accompanied by a signi-ficant remodelling of various parts of the body, with a programmed cell death of different tissues present in nymphal, larval or embryonic forms.

In humans, apoptosis plays a large part in the development of the embryo and the foetus. A known example is provided by the morphogenesis of the hands. In effect, up to an advanced stage of embryonic development, the fingers of the hand are not separated one from the other, but are joined together with a membrane. The hand of a human foetus is thus "webbed" at the start of its formation and the interdigital tissue disappears progressively by apoptosis (D. Duboule).

Apoptosis also plays a part during the development of the nervous system in vertebrates. It concerns both the precursor cells of the germina-tive zone and the mature post-mitotic neurons at the time when they form synaptic contacts. Cell death, at a very early stage of the development of the brain (germinative zone), is an essential step for the morphogenesis and diffe-rentiation of a normal brain. Indeed, if we invalidate in a mouse, by homolo-gous recombination, the genes coding for proteases playing a major role in cell death (caspases), it causes major developmental disorders of the nervous system characterised by an uncontrolled proliferation of neuroblast precur-sors and the formation of a shapeless brain mass devoid of ventricles.

Cell death also takes place during the final stage of brain development, at the level of post-mitotic neurons during synaptogenesis; it has been proposed that those who cannot make contact with their natural target will degenerate! This is probably the case of 50% of the neurons in numerous regions of the nervous system. The explanation lies, in all likelihood, in the fact that neurons enter into apoptosis if they do not receive a retrograde signal (from synaptic endings to the cell body) in the form of diffusible factors expressed by cells that they innervate. To summarise, we can say that apoptosis is a major process, indispensable to the normal development of tissues and organs in invertebrates and vertebrates.

Conversely speaking, the abnormal triggering of cell suicide, which, as we have already emphasised, can be the consequence of a viral infection, is probably responsible for multiple pathologies characterised by significant functional loss, even by the pure disappearance of certain tissues. Included among these pathologies are chronic neurodegenerative diseases such as amyotrophic lateral sclerosis, spinal muscular atrophy, Parkinson's disease, Alzheimer's disease and some retinopathies. The same applies to vascular events such as strokes, consecutive to arterial thrombosis, neurological complications of AIDS, blazing hepatitis and injuries resulting from drug reactions or generated by certain toxins.

II.1.4.5. Apoptosis and Cancer

The "problematic" of the therapeutic approach to cancer can hardly be addressed without taking the apoptotic phenomenon into account. The relationship between the development of cancerous cells and apoptosis is nevertheless complex. Under normal conditions, if there is a cell proliferation independent of exogenous regulating signals and of the process of contact inhibition, etc. (as is the case in "transformed" cells), the associated genetic responses provoke in principle the "self-destruction" of these cells. Conversely, this is probably why the establishment and progression of cancerous foci are accompanied by an inhibition of the process of cell suicide. It is at least one of the steps occurring early in cancer formation.

This inhibition of the apoptotic processes, within and in the neighbourhood of cancerous cells, also represents a significant factor in the development of metastases. In this process of expansion, the cancerous cells are in effect confronted with tissular environments different from those which surrounded them initially.

Radiotherapy and chemotherapy precisely act by stimulating the suicide of the cancerous cells (and not by destroying them, as was thought for some time). Anti-cancerous drugs such as bleomycin or etoposide work in the same way. Yet, the mechanism involved in this activation (or reactivation) of cell death to fight cancer cells is not fully elucidated. Several arguments are in favour of the intervention of a special pathway, the Fas/Fas-L system[21].

Generally speaking, it seems that the anti-cancerous strategies veer either towards the production of molecules capable of activating the suicide of the cancerous cells (molecules which would be specific for each "family" of cancers), or on the contrary, towards the triggering of suicide in the normal cells "around" the tumours, cells which the cancer is exploiting to survive! Among these normal cells figure those from the vascular endothelium. We know in effect that a number of tumours owe their growth to the fact that they "divert", literally speaking, the vascularisation (angiogenesis) to their gain. One of the anti-cancerous strategies, backed by modern pharmacology, lies precisely in the blocking of this tumourous neo-angiogenesis by inducing the cell death of the neo-vessels feeding the tumour.

Recently (A. Kimchi, 2008), other research has been carried out on the mechanisms of apoptosis in relation to cancer. A number of apoptotic genes have been described. This is the case for genes of the DAP family (Death-Associated Proteins). Some of them code for an enzyme of phosphorylation, the DAP kinase (DAP-K). This is a serine-threonine kinase, dependent on the Ca^{++} calmodulin couple. This DAP-K can induce cell death by provoking distortions of the cell or mitochondrial membrane. DAP-K therefore functions like a tumour repressor. In fact, it exercises its antagonistic action at two stages of tumour progression; either at an early phase of oncogenic transformation, or later, at the time of the metastatic phase. It is also striking to note that in numerous human tumours the expression of the corresponding DAP gene is abolished, most often after methylation of its promoter. These facts open the way for new strategies in the fight against cancer.

Another gene, BID, once activated, of inducing the apoptosis of cancer cells, has also been identified. This gene triggers a programme of apoptosis by causing, here again, a permeabilisation of the mitochondrial membrane.

---

21. Fas is a trans-membrane glycoprotein receptor (discovered in 1989) present on the surface of many cells, especially on those of the thymocytes, activated T cells, hepatic, cardiac, renal cells, etc. This receptor is coded by a unique gene (Ch 10, in humans). Fas transmits a cell death signal during its interaction with its ligand, FasL, a member of the TNF superfamily (Tumour necrotic factor). The product of gene p53 (tumour suppressor gene) will activate precisely the production of FasL at the surface of the tumourous cells, Fas[+]. Unfortunately, the majority of tumourous cells are resistant to death induced by anti-cancerous drugs acting in this way. They perhaps owe their resistance to the fact that a FasL[+] cell would be capable of killing the T lymphocytes infiltrating the tumour. Research is underway calling on other ligands than FasL, for example, TNF-α.

Finally, a general mechanism of apoptosis, used in the fight against cancer, has been brought to the fore recently. In fact, it has been observed that when certain receptors are not "occupied" by their ligand, they are capable of inducing a cascade of reactions leading to cell death, which is not the case in the presence of their ligand when their building site is occupied. Among these figure netrine-1 receptors such as DCC and UNC-H. Characteristically, some of these receptors are lacking in numerous human cancers. Such is the case in particular of DCC, whose gene is deleted in 70% of colorectal cancers!

Research aiming at explaining the mechanisms involved in the action of "unoccupied" receptors seems to indicate that they would undergo an early cleavage by the "caspases", and this would cause the unmasking of a peptidic, pre-apoptotic domain, called ADD, which would interact with the DAP-kinase (see above) so as to block its suppressor effect.

II.1.4.6. Molecular mechanisms of apoptosis

The molecular mechanisms taking place in apoptosis, its triggering, its execution or its blockade, have given rise to innumerable studies. These mechanisms are extremely complex due to the involvement of multiple positive or negative interactive loops and of the corresponding reactions in question and also the underlying genetic functioning. Therefore, we can only give here a very schematic outline of the present knowledge regarding this phenomenon. The whole mechanism of apoptosis probably appeared very early in the course of evolution. It is based on the action of a series of genes or gene products either as "effectors" (which some call "executors"), or as inhibitors (also called protectors), or even inhibitors of inhibitors!

Very thorough studies have been conducted on the genetic control of apoptosis in the nematode worm. However, gene homologues have also been found in drosophila, mice and humans. In *C. elegans*, 4 genes, ced-3, ced-4, ced-9 and egl-1, play a dominant role. "ced-3" codes for cd-3, a protein-precursor of a "caspase", which as we recall, is a cysteine-protease, responsible for the fragmentation of various proteinaceous, nuclear and cytoplasmic substrates, thus contributing to the self-destruction of the cell; ced-4 codes for an "activator" of the protein cd-3. In other words, it converts its inactive "precursor" form into an active enzyme (this activation requires the direct interaction of the cd-4 protein with the cd-3). Cd-9, which is what could be called a "protector", prevents cd-4 from activating cd-3! As regards egl-1, it neutralises the protector effect of ced-9. Its activation thus can set in motion the process of cell self-destruction.

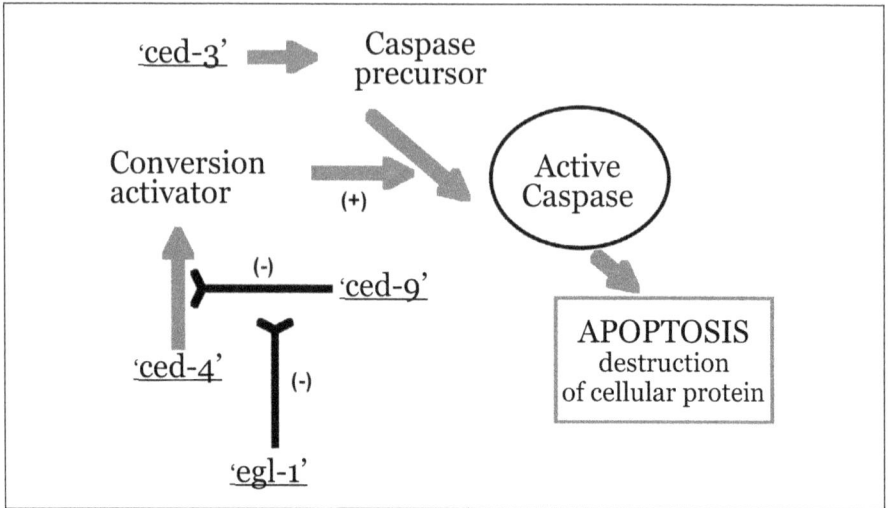

Fig. 6 *Genetic network which controls the balance between cell life and death (apoptosis) in the nematode worm C. elegans*

The "ced-3" gene codes for synthesis of a (inactive) precursor in "caspase" (cysteine protease); "ced-4" manufactures a product which switches on conversion of the precursor to active caspase, but its action is blocked by "ced-9" which is kept "silent" by "egl-1". A signal switching on "egl-1" can trigger a genetic cascade leading to apoptosis of the nematode worm.

This amazing system illustrates the degree of accuracy with which the life and death of every single cell are controlled!

There are, in addition, 6 genes involved, not in monitoring the balance between life and death, like the previous ones, but in the phagocytosis of cells undergoing apoptosis under the action of neighbouring cells.

In the human species, numerous gene counterparts in the "executor-protector" category are known, especially those corresponding to ced-3, ced-4 and egl-1. Homologues of the protector gene egl-1 of *Caenorhabditis* constitute the Bcl-2 and BID family. A homologue of the activator cd-4 called Apaf-1 has also been described. In addition, the existence of antagonists of the protector gene Bcl-2, such as Bax, Bid, etc., have been described. When the Bax gene intervenes, the factors produced by Bcl-2 and its antagonist Bax form an active heterodimer. This has the capability of creating "pores" in the membranes of the endoplasmic reticulum and of the mitochondria.

Deterioration of the mitochondrial membrane, resulting from the neutrali-
sation of Bcl-2, triggers the liberation in the cytosol of internal mitochondrial
proteins. Among these are the cytochrome C, the factor AIF (or apoptosis-
inducing factor) and Smac/Diablo. Once in the cytosol, the cytochrome C or
Smac/Diablo activate the caspases.

As we can see, the mitochondria occupy a strategic place in apoptotic
signalling. To this already quite complex picture, one can add the fact that very
often, caspases (types 3, 6 and 7) are activated by so-called "initiators", some
of which are in their turn activated, by an upstream signal: the interaction of
the Fas receptors or the TNFR (Tumour Necrosis Factor Receptor) with their
respective ligands, the cytokines.

Finally, these signal amplification cascades leading to cellular apoptosis
can themselves be interrupted by various factors, such as the IAP proteins (inhibitor of apoptosis), which block the activity of certain caspases and, furthermore,
the IAP proteins can be inhibited themselves by the Smac/Diablo factors!

One can see with this schematic overview that the mechanisms of
apoptosis (a mechanism depicted as the "clash of opposites", according
to the expression of J. D. Vincent and J. L. Ferry (2005)) has been remarkably conserved during evolution and must constitute one of the principal
solutions to the maintenance of cell autonomy, in the face of varied environmental influences, such as all the external, physical-chemical factors or
the multiple interactions resulting from the vicinity of other cells within
the tissue.

II.1.4.7 Cancers

• *Epidemiological facts*
Cancers belonging to the category of multi-factorial diseases are very
widespread and their prevalence remains high, despite the many research
efforts, prevention and therapeutic strategies to fight them which are deployed
nearly all over the world.

Their appearance dates back to far-off periods of history (obvious signs
of cancer have been found in some Egyptian mummies). They constitute a
very significant factor in the mortality of the world.

• *Cancers in the world and their growing incidence in the developing countries*[22]
If for a long time, people have seemed to ignore the impact of cancer in many countries of the southern hemisphere and especially in the developing countries, the reason for it probably was that the attention was predominantly focused on infectious diseases responsible for all kinds of epidemics, and primordial factors in the limitation of the average life expectancy. We now know that the developing countries are not immune to this scourge.

According to the figures of the WHO and the International Union against Cancer (UICC) (2005), there are in the order of 7 million new cases of cancer per year in the world and in the order of 1.1 million deaths! Cancer kills more people than AIDS, tuberculosis and malaria together and is responsible, for example, for a quarter of the overall deaths in the United States. The percentage of cancer-related deaths is relatively lower in poor countries but in absolute terms, due to the size of populations of the Third World, there are more cancers and deaths from the disease in developing countries than in the developed ones. Between now and 2020, one anticipates that the cancer "toll", in terms of mortality, will exceed 16 million cases, of which 70% will be in countries of a low or average economic index. Worldwide, 45% of cancers occur in people aged above 65 and the number of the elderly is expected to increase 250% by 2050.

The change in lifestyle of people living in countries with weak or average economic income will be one of the important factors, along with the ageing of the population, in the raised incidence of cancer. Among the changes in question, one must take into account the fact that the consumption of tobacco is significantly increasing in developing countries, especially among women. Among the 8 types of cancer considered to be the most dangerous, 6 are in fact due to the consumption of tobacco (especially, and as we all know, lung cancer).

Another cause of the expected increase in the number of cancers is linked to excess weight and obesity, more generally to a sedentary lifestyle (for example, 60% of Americans above 20 years old are overweight or obese).

22.The reader is invited to consult also the excellent article of Joe B. Hardford, director of the International Bureau at the National Cancer Institute (NIH, USA) in "Changing lives, Biovision Alexandria, 2006, Edit : Ismail Serageldin and Ehsan Masood with Mohamed El Faham andAmani Massoud, p. 137 (2007).

In the recent years, the developing world has started to be confronted with the obesity syndrome. As we have repeatedly pointed out, significant portions of the populations of developing countries live in fact in suburban areas and because of the poor conditions of a dietary balance and of changes in lifestyle, they are subject today to diseases which were considered as typical of the rich countries. The rise in obesity is one of these factors.

Recent epidemiological studies regarding cancerous diseases in developing countries have highlighted the situation, especially in the Middle East, where the incidence of cancer is raised but with typologies fairly distinct from one country to another. (It is remarkable that despite the political and economic difficulties of this part of the world, concerted efforts for anti-cancer research and medicine have given rise to the creation of a "scientific-medical consortium", which moreover receives the support of *the National Cancer Institute*.)

The African continent also deserves special attention regarding the epidemiology of cancerous diseases. *"A little more than fifty years ago we considered this continent as devoid of cancerous diseases, since infectious and parasitic diseases dominated the scene and led to such an early mortality that cancer, mainly touching adults, was ignored"*, (Guy Blaudin de Thé, in "Sciences and developing countries, report RST n°1, p. 103 (2006), EDP Science editions").

The vision of the problem would change towards the end of the 1950s, with the now famous observations on cancerology made by the surgeon Denis Burkitt, who worked in Kampala, the capital of Uganda. Struck by the frequency of the occurrence of infant lymphomas, he succeeded in showing that "the particular environment" of the region was a significant factor in the etiology of cancer. In effect, it was established by Burkitt and other epidemiologists that the altitude, temperature and humidity, together with a precarious economy, intervened jointly in the onset of this lymphoma due to the action of the EBV virus[23].

In the second half of the 20th century, as life expectancy in Africa increased, so did the incidence of cancer, and national and international authorities began to be aware of the growing importance of cancer in public health.

The International Centre for Cancer Research, situated in Lyon, has published its available figures concerning the prevalence and nature of cancer

---

23. EBV: Epstein-Barr Virus.

in Africa. It was noted that, although the Anglophone countries (Nigeria, Kenya, Uganda and certain countries of South Africa) in general had at their disposal some records describing the cancer outbreaks for decades, the situation in the field of epidemiology in Francophone African countries was far from being as satisfying. In these latter countries, epidemiological data are, most often, provided by the laboratories of pathological anatomy and they relate especially to solid tumours.

This situation is especially seen in inter-tropical Francophone Africa. It shows a raised frequency of lymphomas and sarcomas. In Gabon, prostate cancer is particularly dominant, as well as cervical and breast cancers. Tropical Africa is characterised by the raised frequency of non-Hodgkins lymphoma and especially by a real "explosion" in Kaposi's sarcoma since the meteoric spread of HIV/AIDS.

In Cameroon, breast cancer comes in first, followed by lymphoma and Kaposi's sarcoma.

On the island of Madagascar, one predominantly notes cancers of the oral cavity (use of bethel) and colorectal cancers, with the prevalence of melanoma and the same types of cancer as in East Africa.

The Anglophone countries of Africa have a preponderance of Kaposi's sarcoma (50% of the adult tumours treated in the hospital in Kampala). According to the studies by de Thé *et al.*, HIV plays a determining role in promoting the oncogenic potential of a latent virus, HHV-8, without, however, reactivating it in 60% of cases.

Finally, in the countries of Mahgreb, smoking promotes cancer of the larynx, the pharynx and the lungs. One also notes (as in China) a high frequency of a specific type of cancer, that of the rhinopharynx (very rare cancers in other regions). This cancer is caused by the EB virus (Epstein-Barr), whose action is empowered by certain foods and by genetic factors.

As can be noted by these facts, cancers are a growing threat to public health in Africa, with some forms identical to those observed in other regions of the world, but also some specific forms (or, in any case, forms encountered in a near explosive fashion on the African continent, as we have seen, such as Kaposi's sarcoma, or Burkitt lymphoma).

In order to help these populations, it is essential to adapt the protocols of chemotherapy to be more effective, but with lower risk of introgenic effects.

This is especially the goal of the international network named INCTR (International Network for Cancer Treatment and Research) whose headquarters are situated at the Pasteur Institute in Brussels.

• *Biology of cancer – oncogenes – suppressor genes – repair system*
    The nature and causes of the cancers capable of developing within most tissues or organs of vertebrates can be extremely varied. We all know that environmental factors, as much as the genetic terrain of individuals, should be taken into account in their appearance. The incidence of such or such a family of cancers therefore varies by way of the geographical distribution of the population, their habitat, their food, etc., and without doubt, also their genetic polymorphism. It would be unrealistic to attempt a clarification, however incomplete, of the innumerable projects dedicated to the epidemiology, physiology, genetics and therapy of cancers. The only ones that could be mentioned here are a few perspectives, some of which are recent. Some aspects of the problem, especially those concerning the "cancer-apoptosis relationship", have been discussed in the preceding sections.

    To say that cancer is a "multi-stage" phenomenon is to say that it is linked to a cumulation of genetic or cellular events, or events of environmental origin, which, when occurring within a given cell, will render it malignant.

    The history of ideas relating to cancer cannot be retraced here: for example, biochemical theories (glycolysis anomalies (Warburg), chemical theories (carcinogenic substances, etc.) mutational theories, hypotheses linked to the loss of cell communication, etc. However, it is clear that with the progress of molecular biology, genetic engineering, and more recently of genomics and cellular immunology, as well as a better knowledge of stem cells and of the mechanisms of apoptosis, biologists have gained a reasonably good view of the molecular and cellular mechanisms which are in play in the malignant transformation, in tumourgenesis and in the appearance of the metastases. The discovery of the oncogenes, at the end of the 1970s, constituted the first serious steps towards the modern genetic study of cancer. It was found, especially in the light of work regarding the neoplastic transformation in mice by viruses such as SV40, Polyoma, etc., that these viruses include in their genome genes (such as the gene coding for the T antigen) which are capable of inducing such a transformation. In 1976, Stehelin, Bishop and Warmus, following research on Rous' sarcoma, a tumour developing in birds, established that the "cancer genes" carried by the transforming virus (like the "Sarc" gene (Src) of the virus responsible for this sarcoma) have their counterpart in the cell itself under an allelic form, C-src. Hence the generalisation about the existence in any normal cell of a family of genetic determinants, called proto-oncogenes (or cellular oncogenes), that play an important role in signalling events exogenous to the cell nucleus, allowing it to divide in pace with its environment. Proto-oncogenes regulate in some way the "social life" of the cell in relation to its environment and when a tissue is formed, make the cells, entering in contact one with another, stop dividing (contact inhibition). We now know that a "normal" eukaryotic cell thus includes in its genome a "network" of around forty genes or more, intervening in this type of regulation. As previously outlined, the first representatives of these have been characterised in birds (Src, myb, myc, erb, etc.) These proto-oncogenes code for different proteins which function by interacting with one another and with the environment. Thus, a "natural" control of the divisions under the influence of external factors (physical, chemical, biological, etc.) is ensured. Among the products of cellular oncogenes figure certain enzymes (e.g. some kinases), transcription factors, membrane receptors, etc. This entire genetic network functions in harmony.

On the other hand, if some of these proto-oncogenes become the site of mutations, whether spontaneous, experimental (irradiation) or even caused by viruses, one can witness the appearance of cancers whose typology and nature will most often depend on that of the affected proto-oncogene. The mutated proto-oncogene becomes an "activated" oncogene; its expression escapes the regulatory signals emitted by other genes in the signalling network as well as the surrounding signals. The cell which is the target of this mutation has now acquired the potential of dividing in an uncontrolled manner.

• *Sentinel genes*
     That being said, the modification of a proto-oncogene only rarely suffices to generate in itself the appearance of a cancerous state. Other muta-tions must take place at other levels of the cell physiology, for there exist naturally occurring, genetically determined barriers. In the early 1970s, a category of gene determinants was discovered, which functions as a "first line of defence" against the de-regulatory effects of the activated oncogenes; they were initially called anti-oncogenes, then "suppressor genes", the best known ones being: Rb (for retinoblastomas), VHL, APC and especially p53, whose genetic modifications are very frequently associated with the manifestations of human cancers. For example, mutations of some of these "sentinel" genes, whether sporadic, UV-induced or consecutive to smoking, etc., are observed in more than 2/3 of colon cancers, in half of lung carcinomas and accompany around fifty varied types of cancers. It has been established that when DNA undergoes injuries under the effects of ionising radiation, or of replication errors, gene p53 is activated. It then triggers the mobilisation of other genes $(GADD_{45}, MDM_2, P21\text{-}WAF1/CIP\text{-}1$ and cyclin G).

P21-WAF1/CIP-1 is an inhibitor of the cellular cycle, at the phase of the G to S transition.

The stopping of the cellular cycle of the damaged cell prevents the propagation of an injury which could have noxious consequences for the tissue or for the entire organism. The damage cell can then enter into apoptosis.

"Sentinel" genes are often specific to different target tissues, for example: APC in the case of colorectal cancers, Rb for retinoblastomas, VHL for kidney cancer, etc.

Besides these sentinel genes (gate-keepers) which act by blocking the division or by triggering the apoptosis of transformed cells, following the mutation of a proto-oncogene, there are other categories of resistance genes, so-called "caretakers", which are involved in the DNA repair systems.

Before expanding on their mode of action, we will discuss some recent work relating to gene p53 and to related genes coding for its isoforms.

Certain authors have pointed out that the mutations of gene p53, generally of the mis-sense type, occur at relatively late stages of the tumour progression. They have reasoned that they could not therefore be responsible for the inactivation of the surveillance system in the early stages of malignant transformation. However, different variants of the protein p53 could possibly be involved. These could be other representative products of the p53 gene family, such as p63 or p73, or some proteins formed by alternative splicing of the gene p53 primary transcript. Such is the case, for example, of p53i2, which includes a copy of intron-2, including a "stop"codon. The product of p53i2 therefore constitutes a truncated version of the p53 protein, named ΔNp53. This truncated protein, as well as the isoforms produced by genes p63 and p73, act as dominant negative factors, which consequently interfere with the action of the wild protein p53 at the early stages of malignant transformation.

Sequence polymorphisms (SNPs) in gene p53 have also been brought to light. For example, PIN-3 is a SNP affecting intron n° 3. Its effect causes a change in the p53/ΔNp53 ratio in favour of the truncated form, which as we have seen, increases the susceptibility to cancer.

• *Repair systems and cancers*

Contrary to the accepted notion, cellular DNA, present in the nucleus and in the mitochondria of the eukaryotic cells, regularly undergoes many chemical modifications (some of which play a possible role in the adaptation of organisms to environmental changes and as motors of evolution, but which, most often, can have a direct or indirect negative effect on the cell). These can be of endogenous origin, and linked to metabolic changes (for example, highly reactive oxygen substances produced during respiration, intermediaries of DNA methylation such as S-Adenosyl methionine), or they can be exogenous (e.g. radiations, sun, UV, etc.).

If we confine ourselves to the endogenous ones, which seem to be the most frequent, they can have a variety of molecular consequences: damaged bases, nucleotidic sites called abasics (having lost the basic component), but still included in a phosphodiester bond, breaks occurring at the level of a single or double strand, etc. These injuries can cause multiple cancers and this is where the repair systems encoded by caretaker genes come into play.

The most frequent spontaneous injuries are the breaks of the phosphodiester skeleton ("single strand" breaks). Then comes the loss of purines or of pyrimidines and the oxidation products (8-oxoG and thymine glycol). In human cells, the number of lesions is estimated at 3000 per cell and per hour. Several dozens of thousands of nucleotides are thus modified, in a constant fashion in the human genome. The preponderant repair method involves the excision of the abnormal base (base excision repair or BER) even if other repair modes are known.

The process of repair by BER mobilises the intervention of several types of enzymes endowed with specific recognition of the damaged bases (for example, DNA glycosylases, of which there are more than a dozen in man). The repair consists of a cut in the N-glycosidic bond (between the base and the deoxyribose), that of the phosphate link upstream by an endonuclease, and finally the complete reconstitution of the nucleotide by a β polymerase. In addition, a protein, XRCC, promotes the coordination of these various events.

It is thought today that mutations of genes of the BER repair pathway (especially implicated in the endogenous lesions of DNA) would contribute to the creation of a "mutator" phenotype which would accelerate carcinogenesis.

Among the DNA glycosylases whose mutations would be the most important in this respect figure the glycosylases UNG, MYH and OGG_1. Epidemiological analyses in man indicate that the inactivation of these DNA glycosylases can lead to a predisposition towards cancerous pathologies (similar observations have been made in mice (S. Boîteux *et al.*, 2006)).

• *Epigenetic factors*[24]

Apart from the proper genetic effects involved in the acquisition of a pre-cancerous state, for example, through mutation or failure to repair, modern oncology has also highlighted the importance of epigenetic factors. These factors do not play a part in altering the DNA sequence in specific genetic sites, but in causing other types of hereditary changes, such as modifications in the methylation state of the CpG islands. As we recall, this short dinucleotide sequence is generally situated at the start of the transcriptional activity of the gene.

When the cytosine is methylated, the neighbouring gene is inactive; correlated with this is the observation that in colorectal cancers, for example, one observes a local hypermethylation of certain CpG sequences (De Murzo *et al.*, 1999), and likewise for other types of cancer (Kuss *et al.*, 1997; Jones and David, 1999). These epigenetic changes would have an effect of annihilating the action of the suppressor genes (for example, Rb, VHL or p^{16ink}-4^a).

The importance of the epigenetic changes in cancerous development has also been established during experiments showing that, if the general level of DNA methylation is reduced to 10% of its normal rate, by introducing a hypomorphic allele of the DNA methyltransferase (the "Dormt" gene), that would be sufficient to trigger the appearance of cancer in mice (Gaudet *et al.*, 2003), probably due to some trigger effect causing chromosomic instability. Studies focusing on human colorectal cancers have also shown that 30 to 40% of patients show a loss of parental impregnation (imprinting) at the level of the IFG_2 gene, following the loss of methylation of the maternal allele IFG_2, both within the tumour itself and in the cells of surrounding intestinal tissues.

Much work has established that the epigenetic changes accompanying the cancerous state included not only a hypermethylation of the CpG islands, in the vicinity of the suppressor genes (such as p53) – which abolished the barrier represented by these genes – but, conversely, a hypomethylation of the CpG islands at the level of genes having a tumour promoter effect (Feinberg and Vogelstein, 1983; Kaneda *et al.*, 2004). Among these "cancer-promoting" genes are the genes BCL_2, MDR_1, HOX_{11}, Myc, c-Ha-Ras, c-Fos, α-FETO-PROTEIN, MASPIN, MELIS, EGRI, SYNUCLEIN GAMMA, etc.

• *Epigenetic control of differentiation in cancer stem cells*

Cancerous tumours, like normal tissue, include specific cells capable of self-renewal ((Levan and Hauschka, 1953; Makiro and Kano, 1953; Sachs

---

24. See also the chapter dedicated to epigenetics.

and Galily, 1955; Hayashi *et al.*, 1974; Kleinschmith and Pierre, 1964; Bonnet and Dick, 1997; Kondo *et al.*, 2004). The presence of stem cells present in tumours has, indeed, many therapeutic implications; it can be understood that the complete eradication of these cells is necessary to avoid recurrence. In addition, because normal differentiated cells can become cancerous, the stem cells found in tumours can come from normal stem cells and from more differentiated cells, reprogrammed as stem cells.

Although in the majority of cases, genetic or epigenetic changes, which are associated with cancers, are characterised by aberrant growth and differentiation, it has been shown that cancerous cells do not always lose their capability of differentiating into mature cells. Such a "return" towards a normal differentiated state can be caused by the addition of cytokines or differentiation agents such as dexamethasone, retinoic acid or cytosine arabinoside. This "redifferentiation" is concomitant with an epigenetic "reprogramming". It can be triggered by inhibitors of the DNA methyltransferases or of the histone deacetylases (Cameron *et al.*, 1999; Wang *et al.*, 1999; Marks *et al.*, 2002; Meuster *et al.*, 2001) and constitutes a new approach for cancer therapy.

Like the normal embryonic or adult stem cells, cancer stem cells are endowed with plasticity in their differentiation potential. For example, we know from the early research by Kleinschmith and Pierce, 1964, and Martin and Evans, 1975, that the Embryonyl carcinoma cells (or EC) derived from testicular teratocarcinomas can be induced (like ES cells) to differentiate into a plurality of cells belonging to the three embryonic layers, and also into neurons, or cells from skeletal or heart muscles. Thus, the cancerous stem cells of leukaemia can differentiate into a variety of normal cell types (blood, brain, colon and liver, respectively). Studies on a line of myeloid leukemic cells using the technique of DNA chips (gene profiling technique) have shown that besides the genes which are normally and preferentially expressed in the hematopoietic tissue, this line was expressing at a high level more than 120 genes, that are preferentially active in the non-hematopoietic tissues, such as neurones, muscles, liver and testicles, a result indicating that the leukemic stem cells are potentially capable of a return towards varied cellular phenotypes.

Speaking more generally, numerous works, impossible to describe in this section (see review by J. Lotem and L. Sachs, *Oncogene*, 25, 7663, 2006) indicate that the chromatin of cancerous stem cells (not only originated from established lines but also from primary cancers) shows transcriptional accessibility vis-à-vis genes normally expressed in tissues other than those at the origin of these cancers. Hence, one can conclude that, in appropriate conditions, the cancerous cells manifest a plasticity in their differentiation, far more extended than had been anticipated. It is interesting to mention in this respect, that the transfer of nuclei coming from medulloblastomas (Li *et al.*, 2003), leukaemia, lymphomas, mammary carcinomas or melanomas (Hocheddingler *et al.*, 2004) into the enucleated ovocytes of a mouse, gives rise to an (apparently) normal embryonic development, at least up to the pre-implantation stage.

Finally, it has been established that the heterologous expression of genes, other than those activated in the tissue of origin of the cancer – whether these genes are proper to the germinal lines (cancer/testis genes or CT genes) or to somatic tissues – can contribute to the development of the cancer itself, by stimulating its viability, growth or metastatic potential. Often, the over-expression of these genes, whose expression is characteristic of tissues distinct from the original tissue of the cancer, confers anti-apoptotic properties, increasing the proliferating capacity or the chromosomic instability.

These properties of heterologous somatic expression by cancer cells can also constitute interesting additions to their diagnosis. Thus, in using the system of genetic probes on solid support (micro-array), it has been established that, among patients suffering from paediatric leukaemia, those with lymphoid leukaemia (ALL) expressed 3 times more non-hematopoietic genes (especially genes of neuronal and testicular phenotypes), than those suffering from myeloid leukaemia (AML). It has been noted that, among the former category of patients, the incidence of leukaemic implication in the nervous system was clearly much higher than in the latter.

No wonder, therefore, that much research in modern oncology is attempting to focus on the properties, and especially the epigenetic plasticity, of the stem cells associated with various types of cancers.

II.2. AGRICULTURE – NUTRITION – FEEDING MANKIND – THE CHALLENGES OF MALNUTRITION – TRANSGENIC PLANTS (DATA, HOPES AND FEARS)

II.2.1. Feeding mankind – Data on the problem and the challenges to be met

Since time immemorial, the problem of access to food in various communities, on whatever scale (family, nation, continent), has been a major preoccupation of humanity. For instance, poor harvests as a triggering factor in the French Revolution of 1789 has been well documented by historians (Leroy-Ladurie, etc.), just as the potato blight played a key role in the mass immigration of Irish populations to the United States. Examples of interdependence between political and cultural developments of nations and food availability are legion. In the 18th century appeared the first theories proposing the regulatory effect of the available food resources on the optimal size of populations (Malthus, etc.). It was only in the middle of the 20th century, however, that economists clearly demonstrated mid and long term repercussions of the accelerated demographic growth of the planet (despite human losses linked to large international conflicts) on a geopolitical but also environmental, sanitary and nutritional scale. We have also seen the progression of a new consciousness of the issue of economic equilibrium between the countries of the Southern and the Northern hemispheres since the disappearance of the colonial empires. Hence the development of terminology describing degrees of development according to a scale going from the poorest countries (so-called "least developed" and "developing" nations) to the "rich" or "developed" nations, with the intermediary status of "emerging" used to designate certain of the most populated countries of the planet (though these terms do not address the large socio-economic disparities concerning the communities or ethnic groups inside each of these countries). Newly conscious of these economic imbalances as factors of political instability and local conflicts, aware of the often irrational exploitation by "multinationals" of natural resources (many of which are located in the Southern hemisphere), and also sensitive to disturbances in the geo-climatic environment and biodiversity, attuned to the pressure on the availability of drinking water, the overall "mentality" of our contemporary world has changed. People have become aware of the necessity of improved management of the shared heritage and resources of all kinds on an international, and even planetary scale. Thus, in the second half of the 20th century, numerous international organisations with specialised vocations (WHO, FAO, etc.) or with general, cultural and economic-political motivations (UNO, UNESCO) were created, as well as a myriad of NGOs mandated for direct local intervention. The world's more advanced countries (economically speaking) have met in "Summits" in order to create large international programmes and to decide on appropriate operational choices. The concept of sustainable development (attributed to the president of the WHO, Ms. Gro-Brutland) was first advanced in the 1980s, its main concern being better management of global assets and heritage ("common goods") for the benefit of all, including future generations. Since the Rio Summit on the environment (1992) a number of international Summits have followed to address the major decennial issues and concerns on the continental and global levels. Among the Summits associated with sustainable development, the one in Johannesburg (World Summit on sustainable development, or WSSD) assumed a particularly symbolic dimension, due to its chronological coincidence with the start of the 21st century. This meeting launched a dozen major initiatives intended to ensure better economic equilibrium, reduce extreme poverty, improve conditions for women, and foster mass education. It also aimed to find remedies for the major damage to the environment. Obviously, the fight against infectious diseases and the fulfilling of essential food requirements appeared as components of this huge concerted effort. To achieve these ambitious goals, the accent was admittedly put on the need for better governance (political, economic and managerial), but also on the hopes raised by recent advances in the domain of science. The International Council for Scientific Unions (ICSU), which had taken an active part in the Johannesburg Summit and whose action is often in line with UNESCO's major programmes (likewise some large federations such as the InterAcademy Panel on International Issues (IAP), or the InterAcademy Council (IAC), conducted a series of studies and made various proposals shortly after the Johannesburg Summit. Some of these were specifically aimed at examining the extent to which recent advances in genetics, genomics and also in biotechnologies, can contribute to fulfilling internationally recommended "millennium objectives", and for this reason focused on the rapid development of sustainable agriculture. We will attempt to show the broad outlines of this approach.

Between 1960 and 2000, the global production of foodstuffs increased in extraordinary proportions due to intensive agriculture linked to an increase in agricultural mechanisation, and the systematic use of pesticides. As we shall see, this led to overabundance in industrialised countries without, nonetheless, solving the food problems of the South. Yet some developing countries, particularly in Asia and Latin America, began to benefit from the modern development of high-yield agriculture and to experience the beginnings of notable economic growth (India), even very rapid growth (communist China, Taiwan, South Korea, Brazil, Mexico), thus partially ensuring their food self-sufficiency.

Throughout this period, we estimate that the world production of cereals doubled and that it has increased *per capita* by around 37%, calorie intake increasing by 35%, while the price of foodstuffs was reduced by around 50% (see the ICSU report: Biotechnology and sustainable development, G. J. Persely, J. Peacock and M. van Montagu, 2002). Among the factors improving agricultural production, one should mention the discovery of "dwarf" varieties of cereals (more robust) and genetic modifications leading to obtaining wheat and rice varieties suitable for high yield cultivation. This dramatic change in the global profile of agriculture has been called the "green revolution" (Borlaug, 1970)[1].

Scientific improvements were not the only reasons for the growth of productivity in global agriculture. Clearly, public health policies, the creation of appropriate institutions, political commitments (notably in Europe), accompanying public and private investments for the development of rural areas, (irrigation, for example) – all these factors together have contributed to reducing, in absolute figures, the number of people in the state of poverty, while ensuring food self-sufficiency, especially in Asia.

This has not, alas, been the case for everyone! Productivity gains in a number of countries, especially on the African continent, but also in some

---

1. According to recent estimates, the green revolution stimulated the global production of food products at the rate of 2.8% per year between 1966 and 1990, while during this same period the world population grew at the annual rate of 2.2%. Unfortunately, this trend would change between 1990 and 1997. While the rate of demographic growth decreased to 1.7% per year, the rate of food production went down to 1.2%. Meanwhile, the number of people suffering from malnutrition reached 850 million...

central Asian countries, and South America, have not followed the general trend seen in the rest of the world.

It was estimated in 2002 (FAO data) that 850 million people world-wide do not have access to sufficient food due to poverty (daily income of less than 2 dollars and in some cases just 1 dollar). Around 60% of these people live in certain regions of South or East Asia and 25% in sub-Saharan Africa. Today, current projections for the global population correspond to an average growth of 75 million human beings per year between 2002 and 2020, with a very large proportion of people living in cities, some of which will be true megalopolises[2].

To feed the planet then, the production of food and grains ought to increase by around 40%, and that of root plants and tubers by nearly 60% by the year 2020 (Pinstrup-Andersen *et al.*, 1995). Likewise, the production of milk and meat will have to double. Of course, these efforts will only be effective if poor families see their average income grow, and agriculture is practiced in a sustainable manner, with the aim of preserving natural resources for the long term, while avoiding the damaging effects linked to old agricultural practices.

Indeed, these traditional practices can have detrimental effects on the environment. They have given rise, for example, to increased salinity in irrigated areas. The sometimes irrational use of pesticides also has certain harmful consequences for human health, ecology and the biological diversity of wildlife.

On a larger scale, one can observe the sad results of huge deforestation practices (for example, in Amazonia and, more recently, in some African countries such as Mozambique). Fishing (see the chapter on biodiversity) has also been practised by some countries to such an extent that numerous commonly consumed species are about to disappear.

─────────────

2. The growth of poor populations, especially in suburban areas, but also the imbalance of dietary regimes connected, generally, with precarious living conditions, in various parts of the world, are leading to the appearance of a set of illnesses or syndromes, today designated by the WHO under the name of dietary transition diseases. These syndromes are expressed, rather paradoxically, through a more rapid increase in obesity in the populations of developing countries than in populations of industrialised ones (in fact, over 75% of women aged around thirty who are noticeably overweight live in countries such as Barbados, Egypt, Malta and South Africa, not just in the United States). This paradox may be explained both through a change in dietary habits and in activities which require heavy physical effort. Obesity is not, moreover, the only manifestation of lifestyle changes in cities; a strong increase in cardiovascular diseases and digestive cancers can also be observed.

However, one of the most serious threats for the coming decades is the shortage of drinking water. At present, agriculture is the human activity which consumes by far the most water. Hence, the importance of new research for an agriculture more economical in water consumption, and the development of plant varieties resistant to drought.

More generally, the limited availability of arable areas on the planet means that new solutions for an agriculture that would be more productive, less polluting and more economical of water and inputs are indeed badly needed.

Lastly, let us note that the use of agricultural varieties in the modern era is restricted, in 90% of cases, to a limited number of representatives of food-producing plants (barely more than 12 crops), and to the use of five main types of livestock as well as two predominant species of fish. Such a restricted number of crops and livestock results in an excessive selectivity, and may lead to the loss of a great number of wild species.

II.2.1.2. A world food crisis – The "return of hunger"

The start of the present millennium has been marked by a new awareness (particularly strong at present) that has mobilised the highest international authorities to address the existence of an extremely serious food crisis in some countries of the globe.

The awareness has been triggered the sudden rise in price of basic foodstuffs in a number of developing or emerging countries. As Pierre Jacquet, head economist of the French Agency for Development wrote (*Sciences au Sud – le journal de l'IRD*, no. 44, April-May-June 2008), "we are seeing a real 'return to hunger', attested by the 60% rise in one year of the food price index of the FAO and the explosion of prices of cereals and dairy products." This situation incited numerous reactions, comparable in certain cases to genuine revolts, some of which have even toppled governments. During the first half of 2008, there were around thirty "hunger riots", which gave rise to an international meeting (on 3-5 June 2008) on food security at the FAO headquarters.

Economists, agro-ecologists and specialists in nutrition issues vary on their estimation of the causes of this worsening situation, but concur that the main ones are:

- A slowdown in agricultural production, to the point that some countries, formerly exporters, are now obliged to import large volumes of basic foodstuffs. The result is often a very pronounced reduction of the stocks that normally enabled these countries to cope with brutal decreases in local production following extreme situations of a natural origin (drought, etc.) or following conflicts;

- A pronounced falling-off of investments in the agricultural sector and the fact that external aid is more often focused on improving large infrastructures than directly helping with local foodstuff production. It is probably significant to note in this regard (see the INRA file, no. 5, June 2008, "Penser la recherche agronomique au niveau mondial") that although the World Bank had not dedicated an annual report to agriculture since 1982, in 2008 it replaced agriculture among the priorities for development; it has insisted since then on the necessity of supporting small-scale agriculture as well as rural employment. Yet, according to the OCDE, the proportion of development aid received by agriculture fell from 11.5% (1984-85) to 3.4% (2004-2005) ;

- The striking increase in energy costs (especially oil) ;

- The demographic expansion and the increasing food requirements in emerging countries (with an increase in meat consumption);

- The increase in the demand for "non-food" agricultural products (i.e. the development of biofuels) ;

- The recent World Summit in Rome (FAO) made a commitment to ensure that global hunger shows a 50% reduction between now and 2015, and that the amount of urgent food aid promised by the participating countries during this Summit reaches 6.5 billion dollars.

The present crisis has contributed to expanding reflection on the middle- and long-term perspectives in the domain of agriculture and nutrition at the global scale. Numerous experts estimate that, to feed a global population likely to reach 9 to 9.5 billion inhabitants between now and 2050, it will be necessary to double cereal production and triple the supply of inputs linked to nitrogen requirements, as well as to increase the use of water by about 90%, not to mention the necessity of putting large geographical surfaces under cultivation, often to the detriment of the environment.

Generally speaking, following the April 2008 publication of the conclusions of the International Assessment of Agricultural Knowledge, Science and Technology for Development (IAASTD), a group placed under the aegis of the World Bank, the FAO, the UN and around sixty states, one must "rethink agronomical research on a global scale". Science and biotechnologies can certainly contribute towards the fight against poverty and hunger. However, this will only be successful if, from now on, improved "synergy between biology, ecology and social sciences takes shape, so as to take into account better the diversity of agricultural situations and their vulnerability" (INRA file, no. 5, June 2008). Keeping in mind this necessity and this new context, we will comment below on some of the contributions of biology and biotechnologies to the agricultural and food-processing sectors.

II.2.2. Contributions of genomics

It is now necessary to get a general view of how the developments of contemporary biology, especially in the domain of genetics, can contribute to meeting the main challenges imposed by a sustainable agriculture.

We are led to examine what molecular genetics, as well as the new technologies associated with it, can offer now or in the future.

To begin with, one can observe that great advances have been made in the knowledge of the physiological functioning of plants, their response to the environment and their mechanisms of resistance to pathogens. This was an indispensable prerequisite to any intervention focusing on concrete solutions. This knowledge ought to lead to a more sensible choice of characteristics to be selected during crossbreeding programmes, with the aim of improving the productivity of food-producing crops, trees, and even animal species such as fish.

Genetics and its recent developments (genomics, etc.) provide, in the first instance, molecular markers for the characterisation and preservation of good agricultural varieties. Hence, the diagnosis of phytopathogens is facilitated, as is improved knowledge of defence mechanisms and the protection of plants against the diseases to which they are vulnerable. The development of vaccination approaches is included in these techniques.

The direct contribution of molecular genetics to agriculture can be described briefly as a two-fold approach involving:
a) improved efficiency of current agricultural practices ;
b) defining new agricultural options with the introduction of transgenetic strains modified for one or several specific physiological traits.

Previous chapters have outlined the principal approaches inherent in structural and functional genomics, and described some of the results from these approaches concerning plants. Important observations have been made, especially regarding cereals (wheat, rice, maize), and the study of certain genes of interest (growth, strength, resistance to pathogens, nutritional quality, etc.) has progressed considerably. Such research benefits from data relating to the synteny phenomenon. Indeed, these three major cereals not only share a large number of genes among themselves, but also with other wild and cultivated plants. Furthermore, the new technologies, based on the determination of the transcriptomes and proteomes, applicable to plant species which can be cultivated, greatly facilitate the analysis of their responses to changes of environment (e.g. abiotic or biotic stresses), as well as the characterisation (through proteomics) of new proteins of interest, and the improved knowledge of metabolic routes.

II.2.3. TRANSGENIC PLANTS – SOME GENERAL DATA

The genes of interest which have been well described by genomics, and which can possibly use transgenesis for developing new properties concern primarily those involved in the resistance of plants to drought and salinity or to biotic stress, (particularly viruses or insects), as well as other important genes for increasing the preservation (storage) of food-producing plants, and their nutritional value.

While conventional methods, such resorting to crossbreeding and the selection of cultivars, have enabled important progress to be made in the acquisition of useful characteristics, biotechnological procedures, especially selection "assisted" by molecular markers and transgenesis procedures, are gradually assuming greater importance for agriculture on the global scale.

The years 2005 marked the 10th anniversary of the commercialisation of genetically modified organisms (GMOs), which some also call "biotechnological varieties". In 2005, plantations of transgenic varieties were estimated (cumulatively over 10 years) at 400 million hectares (in 21 countries of, admittedly, very uneven economic importance). The global surface of GMO plants approved by consumer countries was estimated at 90 million hectares for the year 2005 (compared with 81 million hectares in 2004 and 52.6 million in 2001). In 2005, the United States, followed by Argentina, Brazil, Canada and China were still the main countries producing genetically modified plants, with a global production of 49.8 million hectares for the United States alone. Among European Union nations that are growing (to a lesser degree) transgenic plants (especially genetically modified maize), figure today: Spain, Germany, France, Portugal and the Czech Republic. Finally, of the 21 countries of the globe that adopted transgenesis procedures in agriculture, 11 of them are developing countries and 10 are industrialised. For the total of 90 million hectares sown with GMO plants in 2005, the share of developing countries was globally estimated at 33.9 million hectares.

Despite the strong opposition that continues to be expressed against the use of GMOs, especially in Europe, and notwithstanding the requirements of scientific objectivity that demand that both the risks and the advantages of this new technology be assessed, recent estimates show that accumulated economic benefits (for the period 1996-2004) do emerge. They are equivalent to 27 billion dollars, of which 15 billion concerned developing countries, and 12 billion concerned industrialised countries.

Likewise, the reduction in the use of chemical pesticides between 1996 and 2004 was estimated at 172,500 megatons!

Other data concerning the economic status of people growing transgenic varieties indicates that in 2005, for example, one could count approximately 8.5 million farmers (in 21 producer countries), of whom 7.7 million were described as having low resources (6.4 million in China, 1 million in India).

Transgenic soybeans continued, in 2005, to hold first place among genetically modified plants, and covered a total surface area equivalent to 54.4 million hectares, followed by transgenic maize (21.2 million hectares), cotton (9.8 million) and canola (4.6 million).

• *Principal types of modification introduced by plant transgenesis with agricultural aims*
Among the traits most frequently transferred to transgenic plants (2005 reference), one notes firstly tolerance to herbicides, followed by resistance to insects. In 2005, tolerance to herbicides transmitted genetically in plants such as soybean, maize, canola and cotton concerned 63.7 million hectares (71% of the total of surfaces occupied by transgenic plants). Among the other important traits which GMO producers endeavour to transmit are resistance to drought and salinity, as well as resistance to pests.

• *Drought and salinity*
Agriculture depends increasingly on irrigation. According to Brian Johnson, irrigated areas represented, in 1960, 10% of cultivated surfaces worldwide. This proportion has gone up to 20% today, as irrigated farming contributes at least 40% to total agricultural production. The most serious concern is that irrigation of farmland uses two thirds of the water consumed worldwide, while urban agglomerations consume one tenth, and industry around one fifth of worldwide water resources (the development of emerging countries will, however, increase these last two factors).

The demand for water for irrigation is not only growing regularly, but drought is also increasing in frequency and intensity. One observes that areas that were once cultivated have become unproductive due to recurring drought and the reduction of rainfall. This situation is particularly critical in Pakistan, in the Indus valley, where arable land has experienced increased salinity (increase in deposits of chlorides and sulphates, see FAO, 2003). These phenomena of recurring drought and saline deposits are now frequent in the Middle East, Central Asia and South America, as well as in certain parts of Africa and Australia (FAO report, 1998).

There are various possible responses to the harmful effects of drought and salinity in plants: some are based on growing naturally resistant plant varieties; others on the implementation of genetically modified species.

As regards the recourse to naturally resistant plants, we can cite the example of rice and wheat tolerant to salt and sodium ions, and which are cultivated in India. These varieties were obtained by crossbreeding certain very old representatives of these cereals, which showed a natural tolerance, to high-yield dwarf varieties now commonly used in Asia. Another illustration, based on this type of crossbreeding, concerns a variety of maize whose yield on dry soil is nearly 50% greater than that of traditional maize (CIMMYT, 2003).

However, it is also possible to resort to gene transfer. Such genes have been characterised in numerous plants, including such a model organism as *Arabidopsis thaliana*, as well as other plants not used in agriculture, such as xenophyte plants, and even in some bacteria.

The discovery of the genes of the Dreb series in *Arabidopsis* (1990) aroused great interest when it was observed that these genes could lead, after transfer into plants of agricultural interest, to good tolerance to drought and salinity. So the Dreb 1-A gene was transferred into varieties of wheat, which then demonstrated a prolonged resistance to drought and revealed themselves to be much less demanding of water (CIMMYT, 2004). Likewise, the HVA-1 gene, of agricultural interest, derived from barley, proved to be beneficial in growing plants in arid areas (AGERI, 2005). *E. coli* genes encoding the synthesis of trehalose may give rice double resistance to drought and salinity (Su *et al.*, 1998; Garg *et al.*, 2002). Although these diverse approaches are certainly promising, we ought to question the effect these transgenic varieties are likely to have on the environment (ecosystemic equilibrium).

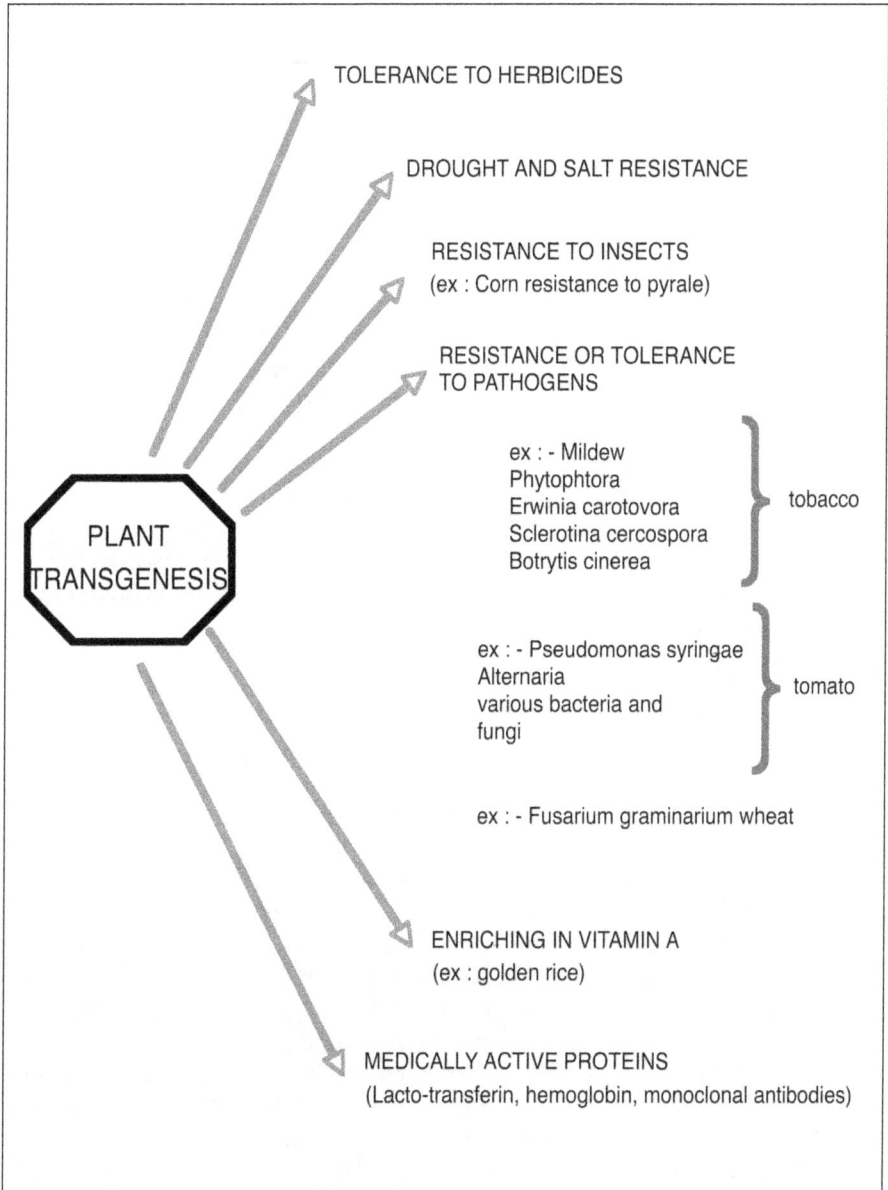

Fig. 7 *Main types of modifications made during plant transgenesis for agricultural or other purposes*

See the chapter on transgenic plants.
See also Biofutur "Plantes à tout faire", 242:16 (2001), ibid. p. 24.

• *Other characteristics*

Apart from resistance to drought and salinity, bio-agronomists have also been attentive to other characteristics likely to be provided by genetic modification. These are resistance to pests, diseases, or herbicides, and also the ability to produce proteins with therapeutic effects, and, of course, the general improvement of agricultural productivity.

Several plants of agricultural interest have therefore been rendered resistant to certain herbicides, and chemical substances generally used for the destruction of weeds. Indeed, this was one of the first physiological traits that the biotechnological industries tried to induce by transgenesis. Among the herbicides in question, let us cite: glyphosphate (better known as Roundup®), phosphinothricine (Basta®), and glufosinate (Liberty®). In 2005, the characteristic of genetically-induced tolerance to herbicides, as used in plants as varied as soybean, maize, canola and cotton, corresponded to an occupation of 63.7 million hectares of cultivable fields (71% of all transgenic plants).

Generally, the potential yield of crops is limited by diseases and pests. For instance, it has been established that in the case of cereals, such losses of annual yield may exceed 25%! In 1998, Africa lost 60% of its cassava harvest due to devastation caused by the African cassava mosaic virus. Reductions in yield due to viral attacks are also frequent in Africa when farming sweet potatoes. Another major objective of modern biotechnological strategies is battle against insects: in 2005, over 10 million hectares were concerned. A frequently used strategy consists of transferring to plants (thanks to the "plasmid" *from Agrobacterium tumefaciens*, in accordance with the technique discovered in 1987 by Belgian researchers, J. Schell and M. van Montagu)[3], a fragment of DNA from a bacterium of the soil, *Bacillus thuringiensis*. This fragment actually contains a gene, the Bt gene, coding for a protein normally accumulated in the spores of the bacillus in the form of a so-called crystal which has a powerful toxic effect on insect larvae. One of the most common applications of this technology focuses on the protection of Bt maize against the European corn borer, a pest whose larvae are particularly harmful. However, many other plants of widespread agricultural use, such as rice and soybean, have thus been rendered resistant to some of their most vicious pests. This is also the case with cotton. Farming transgenic cotton containing Bt gene has resulted in a reduction of over 75% in the use of pesticides, particularly in China, where

---

3. There is an alternative method for the transfer of the Bt gene (and for other examples of transgenesis in plants). This method, discovered over fifteen years ago, is called "biolistic". It consists of "bombarding" the plant cells with micro-particles of tungsten containing a DNA carrier of the transgene.

these agents were traditionally sprayed by hand over the harvests (400 or 500 farmers appear to have died following acute poisoning from pesticides).

Often, genetic modification is not limited to action upon a single gene (previously purified by genetic engineering) but rather concerns several genes (staged gene transfer). For example, maize or cotton plants carrying both the Bt gene for resistance to insects and one or more gene(s) for resistance to herbicides are presently commercialized.

Apart from the Bt genes coding for the toxin *B. thuringensis*, other genes which can transfer the capacity of resistance to insects are beginning to be studied, such as the genes coding for protease and α-amylase inhibitors, or for lectines. The aim, in this case, is to block the digestion of proteins and starch by predatory insects.

In addition to the production of maize and colza rendered resistant to insects, transgenesis has also been implemented to induce resistance to numerous pathogenic agents in plant species of agricultural or other interest (see *Biofutur*, no. 242, M. Durand, Tardiff and T. Candresse, March 2004). Likewise, tobacco (*Nicotiana benthamiane*) has been "modified" by two genes, RPW8 and RPW82, isolated from *Arabidopsis*, with acquired resistance to porous mildew. Another example: the transfer of BCl2 and BCl-X genes, which are anti-apoptotic genes of human origin, prevents the triggering of programmed cell death in tobacco plants that normally would follow fungal infections produced by *Botritis cinerea* or the action of the TSWW virus. Another pheromone gene, OHL, confers to tobacco plants some protection from infection by *Erwinia carotovora*. Other transgenes coding for peptides with anti-microbial activity are able to induce broad-spectrum resistance to bacteria, fungi and insects in tobacco plants. Finally, the "RNA-interference" mechanism has also come into play by transferring fragments of a RNA virus.

The tomato was shown to express a wide resistance to various pathogens (bacteria, viruses, fungi) through over-expression of the Prf gene. The gene of the p35 baculovirus, an enzyme inhibitor playing a part in programmed cell death, prevents the appearance of this phenomenon in the tomato infested by certain fungal or bacterial species.

Another illustration of the possibilities of protection afforded by transgenesis involves wheat. In this case, the transfer of the TriR gene codes for an enzyme that disactivates the mycotoxin *Fusarium graminearum*. As for rice, Japanese scientists (Hayakawa *et al.*, 1992) have introduced, in diverse varieties of this cereal, a "rice strip virus" gene that codes the viral coat protein and obtains protection for this altered rice against numerous viral infections. Similar results have been obtained by using the same transgenic strategy in tomatoes and tobacco.

Lastly, diverse transgenic approaches have enabled the delay of leaf ageing and the breakdown of the "Rubisco" protein, allowing a more sustained chlorophyllous activity of the canopy.

Most characteristics that define food quality are dependent on, or controlled by, more than one single gene. Such is the case for taste, smell, colour, nutritional value and other determining properties. These properties are actually the result of complex biochemical reactions. Likewise, a plant's harvest yield is effected by various characteristics associated with its development, flowering, etc. The same goes for a number of stress-resistance capacities. We know, for example, that several genes play a part in the resistance of crops to fungal infections. The response to drought (described above) involves relatively well-known metabolic changes (in the case of sorghum), which enables the plant to reduce its water consumption.

• *Overall physiology – Nutritional value*
    It would seem reasonable, based on the above information, that future GMO-related technology should increasingly concern the manipulation of properties and overall physiological characteristics of cultivable plants by resorting to the transfer of multiple genes. It would be expedient to look toward the introduction of a new mechanisms of biosynthesis relating to key health component (for instance, vitamin content), to an original process of detoxification of the environment (bioremediation), to the appearance of new horticultural properties, or even to a better yield. Of course, these objectives may also be achieved through successive crossbreeding techniques; but this more traditional approach often demands considerable time investment. Thus, it could be worthwhile for GMO specialists (and users) to focus perhaps on a strategy of combined gene transformation (gene stocking strategy).

It is possible, on occasion, that the introduction of a single gene may nonetheless suffice to bring about a major metabolic change, modifying the plant's overall physiology. For example, concerning the increase in agricultural productivity, certain research projects aim to increase the efficiency of photosynthesis in cereal plants by improving the management of water loss through the leaves, or by regulating the opening and closure processes of stomata (Mann, 1999 quoted in "Changing life", *biovision*, Alexandria 2006, p. 251). Other researchers are attempting to modify the photosynthesis of rice, trying to substitute the metabolic route in C3, through a route in C4, thanks to the introduction of a cloned maize gene playing a part in the metabolism of starch C4 precursors (Ku *et al.*, 1999; Matsuoka *et al.*, 2001). Other attempts to improve agricultural productivity involve the accumulation of starch itself; for the biosynthesis of this polysaccharide plays a central role in the plant's metabolism (formation of reserves in seeds, roots, tubers and fruits). This is why some research is devoted to improving starch storage in the organs of plants by increasing the expression of the gene coding for ADP glucose phosphorylase (gene ADPGPP) (Kirshore, 1994). The starch and dry material content of potato tubers was shown to noticeably increase after transfer of the gene glgC16 *from E. coli*, which codes specifically for this enzyme (Stark *et al.*, 1992). However, these approaches still remain in the experimental phase, and have not yet been used in agricultural applications. Presently, under the same heading as "metabolic conversions", we can also include the emergence of GMO plants destined for the production of molecules with therapeutic effects. Here, the goal is to add a beneficial action for human health (medications, vaccines, antibodies, human proteins, vitamins) to the nutritional value of the plant. This could be described as a "double-effect transgenesis".

• *Plant transgenesis and health*
    A particularly interesting illustration of this double-effect transgenesis is the production of what has been called "yellow rice", enriched with β-carotene, a precursor of vitamin A. By transforming the rice by genes from diverse biological sources (staged gene transfer), scientists have succeeded in obtaining varieties of rice whose yellow colour is due to the accumulation of β-carotene. This is a good way to combat vitamin A deficiency, which affects a number of children living in countries where rice is the near-exclusive dietary staple, such as Vietnam, Laos, Cambodia, Nepal, Bangladesh and India. This deficiency is quite pronounced and always serious. One of its consequences is early blindness, which affects a large number of children; it also causes problems with development and the immune response to infections.

    Several teams tackled the problem of how to obtain a genetically modified rice enriched in β-carotene (Ye *et al.*, 2002). Two genes, one from a flower (the daffodil) and the other from a bacterium (*Erwinia uredovora*) were introduced into a model rice variety, Taipei 309. Some of the fertile plants were used to genetically transfer the carotene biosynthetic pathway into other varieties. However, Taipei 309 rice is no longer cultivated due to its low crop yield. Another team (Hoa *et al.*, 2003) introduced the same genes into strains of Indica and Japonica rice.

    Besides the attempts to strengthen the vitamin content of agricultural cultivars, one should also mention the approaches aiming to increase the content of lysine, an essential amino acid, into maize, soybean and colza. This was achieved by transferring two genes: gene dapA, coding for the precursor of lysine (dihydrodipicolinic acid or DHPHS), a gene from *Corynebacterium*, and gene lys-C from *E. coli* coding for an *aspartokinase* (AK), an enzyme involved in lysine biosysthesis. The lysine content of these genetically modified plants was found to increase about fivefold.

Finally, certain pharmaceutical industries have placed hope in the use of plant transgenesis to obtain proteins with therapeutic effects (enzymes, antigens, antibodies, human proteins). The significance of this approach may lie in the fact that for some developing countries, access to vaccination is not always guaranteed and it is important to fight both malnutrition and certain viral epidemics. One example is the genetically modified banana, expressing the antigen Hbs, which, as we know, has a vaccinating effect against hepatitis B. Generally speaking, transgenic plants could have an important future in the production of medicinal proteins. For example, tobacco is used in the production of lactotransferine and human haemoglobin; a number of plants may be genetically modified for the production of monoclonal antibodies. Let us mention, in this respect, the success obtained by a Cuban team which has produced from tobacco plants an antibody directed against the antigen Hbs (with a maximum output of 25 mg of antibodies per kg of biomass). This antibody can also serve, for its part, as an immuno-purificator agent of the corresponding antigen.

The main interest in producing proteins with therapeutic effect through plant transgenesis arises from the potential yield of cultures on large surfaces; additionally, there is the benefit of lower risk of viral transmission, as compared to animal products, to humans.

Finally, besides the hopes that the pharmaceutical industry has placed in its recourse to transgenic plants, it seems appropriate to mention those that are expressed by the chemical industry. Transgenic plants will, in fact, be able to become potential biological factories able to manufacture chemical products with industrial impact, as some achievements concerning the production of lubricants, perfumes and flavourings[4] already indicate.

---

4. Here, however, the same ethical problems in terms of the food-processing industry as those with the production of first-generation biofuels may appear. If, however, the producers were dealing with large-scale cultures...

• *Hopes – Reservations – Potential risks*

It is clear from the numerous examples illustrating the existing or potential properties of transgenic plants, that the rapid development of new techniques leading to the modification of numerous plant organisms by transgenesis has a growing significance, and it is accompanied by often considerable investments in biotechnologies. Despite the accomplished or expected progress from pure research in genetics, biotechnologies (taken in the broad sense) and more particularly, the production and use of GMOs, continue to provoke loud controversy and to arouse public anxiety, especially in Europe. It seems that feelings for or against GMOs depend less on scientific data than on fear concerning what some consider to be a systematic "commodification" of nature. In this respect, if the public generally puts its trust in scientists, it often expresses very strong reservations concerning the other parties such as industries, governments, administrations (from "GMO, the research challenge', an information document from the French Ministry of Research, 2001)

The public's preoccupations concerning the applications of biotechnology, particularly with regard to GMOs, can be envisaged from four different angles. They may be of an ethical, socio-economic, medical,or environmental nature.

As regards the ethical dimension of the problem, the most common attitude focuses on the propensity to transgress natural laws by crossing biological barriers, fulfilling the Cartesian dream which wanted to make man "the master and possessor of nature...". Here, we touch on what some consider to be the Promethean aspect of an unlimited exploitation of the living world.

A few comments are required here. The first falls within a scientific observation: in nature, gene transfer processes are the rule; mainly in the microbial world (plasmids, transposons), but such is also the case with plants. Secondly, the classic procedures of hybridisation and crossbreeding, used since Neolithic times to feed man and beast, have also modified the soil and often resulted in the loss of naturally occurring wild creatures, well before the emergence of biotechnologies! However, it is true that they did not result in the mixing of genes from one biological kingdom to another as is the case for transgenic interventions.

Finally, if countries with low resources, in particular developing countries, succeed thanks to biotechnologies in reaching a better food balance, and at least partly alleviating the North-South disparities as regards quality of life (the timescale is plausible but still very remote), then morality, like ethics, will have won!

Concerns of a socio-economic nature are in keeping with the serious economic risks that multinational agricultural industry might represent for the continuation of traditional agriculture, and even more so, for the self-subsistence of small farmers. The latter, if they are faced with a "standard" transgenic production on a very large scale, sold at low prices and with a long shelf-life, will soon be deprived of their source of income. This is not an unfounded misgiving. It may be all the more justified since intellectual property law for obtaining transgenic cultivars has rendered traditional agriculture dependent on large companies that hold patents, for example, of modified seeds. Here, we recall the (justified) bitterness of debates about the presence of the "Terminator" gene in the seed of the first GMOs marketed in the United States. The introduction of this gene had been designed to block the germination of grains from the cultivated plant, which forced the grower to re-purchase seeds from the manufacturer every year! This technique is no longer in use, but dependency as regards patents remains.

As regards the safety of foods and human health, issues have been raised relating to the toxicity of transgenic foods, the allergenic potential of artificially synthesised proteins, as well as the nutritional value of these foods.

For the moment, it does not appear that significant effects on human health have been reported. We should recall that some very high consumption products, like those derived from transgenic maize (and transgenic maize itself) have been used for over a decade in the United States, Canada and China without showing any harmful effects specifically attributable to the transgenic nature of these products. Nonetheless, it is correct to emphasise that we probably do not have sufficient visibility to be able to dismiss any risk. Drawing inspiration from the safety-first principle, the European Union has decreed various directives (directives 90-219, revised in 1998 (new directives 98/81/CE) and 90-220 stating the criteria of evaluation of the risks of GMOs to the environment and health, directive 2001/8/CE modifying directive 90-220) on the regulation regarding "new foods" and labelling, and directive 98-44 concerning legal protection of biotechnological inventions. The legislation is therefore permanently evolving. As we can see, labelling of the products put on the market and their traceability (of the successive stages of their cultivation, processing, distribution) have been recommended in Europe.

Because transgenic products appear to threaten the environment to various degrees, misgivings about their use are most frequently evoked by ecologists, although the general public is also attentive to such arguments.

Among the potential ecological risks identified there is the possibility of a dissemination of the "transgene" from the genetically modified variety to its non-transgenic counterpart, as well as to related wild plants in neighbouring fields. One should recall that, at the beginnings of plant transgenesis, GMO opponents had already noted the risk of dissemination of a marker of resistance to antibiotics (which was used for some time by the first GMO plant producers to verify the effective integration of the transgenic of interest in the recipient plant and to select the transformed cells). However, this technique for selecting transgenic plants has been abandoned. Another frequently mentioned risk concerns an undesirable genetic flow conveying a gene of resistance to herbicides, from a transgenic plant species to a wild species. Potential ecological risks may also be associated with the spread of the Bt gene, which elicits in the transgenic plant resistance to pests. Specifically, it has been feared that populations of insects resistant to the Bt gene product would develop and proliferate when they are exposed to the transgenic plant. To reduce this risk, it is presently recommended to cultivate, near the modified maize or cotton fields, some areas serving as kinds of "insect sanctuaries", formed by cultures from the same plant species that has not been transformed. This precaution which seems henceforth to have been generally adopted is all the more useful since living species, such as birds or butterflies, are also to be taken into account in this ecosystem. If it is true that certain precautions then seem justified for protecting the environment from possible harmful effects caused by transgenic plants, it is no less correct to acknowledge that the latter may just as well have beneficial effects on the environment. We have already pointed out, for example, that recourse to some transgenic entities enables a drastic reduction of the use of pesticides, just as the FAO emphasises regularly.

Potentially positive effects that genetically modified plants could have on reducing the impact of agriculture on the environment concern the production of biodegradable plastic and coloured cotton (to reduce dependency on synthetic colouring agents). Presently, one can already observe that the genetic modifications to trees through transgenesis has led to a 50% reduction in their lignin production and to an average 15% increase in cellulose production, which finally results in a greater relative quantity of "pulp" for the same weight of wood. As transgenic trees are generally 25 to 30% larger than non-modified species, this might represent a considerable gain in the manufacturing output of wood pulp. What is missing, admittedly, is the intellectual distance to assess the development of these transgenic forests.

In the same vein, plant transgenesis has enabled the obtaining, at least experimentally, of a wide variety of plants useable in bioremediation, by rendering them able to absorb various metals such as aluminium, copper, mercury and cadmium from contaminated soils.

To summarise, if the general opinion of the specialists is that we should undertake to produce more foods from more restricted arable or cultivable land and rely on a lower degree of irrigation, less labour, lower consumption of energy and agrochemical agents in order to avoid famines and malnutrition in the present century, then we will have to enter deliberately into what Swaminathan has called "the evergreen revolution" (2000). This means that from now on it will be necessary to exploit to the maximum the recent scientific advances made in the domain of the biology of plants. This will be made possible by a wise, rational recourse to genomics, transgenesis, etc., their objective being to facilitate the production of varieties which harbour great potential as regards growth yield, resistance to diseases and to insects and tolerance to stresses such as drought, salinity or unfavourable temperature conditions (Sasson and Elliott, 2004). This is why biotechnology, while controlled by local and international political measures, has been and will continue to be an essential approach for trying to resolve the problems of hunger and malnutrition (Khush and Ma, 2004).

However, changes, even these necessary upheavals, will doubtless never be totally free of direct or indirect risks, just as we have mentioned. From which comes the necessity for our society, on both national and international levels, to be able to assess the appropriate measures with full knowledge of the facts. This entails a dialogue on several levels: biologists, farmers, physicians, nutritionists, sociologists and, of course, manufacturers, politicians and representatives of civil society.

II.2.4. Livestock[5] as a major component of human nutrition – environmental effects and perspectives

Animal farming constitutes almost 40% of the gross domestic product for agriculture (in the widest sense of this term), on the global level, and furnishes a third of all proteins consumed by human beings. Thus, it constitutes a major component the activity that humans devote to normal nutrition, and may even provide a remedy for undernourishment. Its impact on living conditions, particularly in developing countries, is considerable. It should be remembered that it employs 1,3 billion people and allows 1 billion of the truly poor to subsist!

---

5. «Livestock's long shadow – Environmental issues and options (LEAD)» (2006).

Global demographic growth and the rise of average revenues have initiated a rapid rise in the demand for meat, milk and eggs in the past decades. In developing countries, the consumption of meat per person is estimated at 26.7 kg/year, with values as low as 12.3 kg/year in sub-Saharan Africa. On the other hand, these values reach an average of 80 kg/year in developed nations, with peaks of 132.5 and 160 kg/year for the United States or Oceania respectively.

A worldwide scarcity of milk and meat will probably increase as a result of the amelioration of the economic status of numerous emerging and developing nations. According to the FAO, production of meat ought to be multiplied by 1.75 in developing countries between now and 2020, while developed nations need to increase it by 1.15, and milk production must be multiplied 2.0 in developing countries, and 1.05 in developed ones.

Moreover, economic development has happened alongside with intensive urbanization throughout the world, resulting in a displacement of animal farming from the countryside to urban and peri-urban zones. Such production is moving closer to consumers, and is at the origin of a considerable growth of the agro-alimentary industry. Thus, 80% of growth in this sector is now caused by industrial systems.

The need to resort to mass production has caused a redistribution of livestock species. For example, one observes increase in «industrial production» of pigs and poultry, and a certain stagnation of production of cattle or sheep and goats. We also note that another important consequence of the development of animal farming on a global scale is its impact on the environment.

● *Research*
Numerous studies have been devoted to the selection and reproduction of animals. The selection of sturdier stock, specifically animals less sensitive to infectious disease, is beginning to benefit from a better understanding of the genetic heritage of the species in which the selection process is underway. Knowledge about genomes has already been obtained regarding 4 species of particular agronomic interest: cattle, pigs, chicken and rainbow trout. Several genes controlling resistance to disease have been identified in fish. Genes coding for proteins implicated in the fertilizing capacities of spermatozoids (ovine, porcine, equine) have also been characterized (INRA : letter n° 14, May 2006). Transgenesis has been practiced in livestock species for many years (cf. report from the CADAS «Les techniques de transgénèse en agriculture», n° 2, October 1993, Tec et Doc), although until now the technique has not led to any notable applications for agronomy.

On the other hand, thanks to molecular techniques, knowledge and diagnosis of the epidemiology of infectious diseases in animals (PPCB, African horse sickness, bluetongue, foot and mouth disease, etc.) has greatly improved over the past ten years. Thus, in areas where adequate equipment is available, PCR (*polymerase chain reaction*) technology is particularly adapted in that it can, in certain cases, be effected upon dried cultures. This makes it possible to eschew cumbersome temperature controlled supply chains to conserve the samples meant to identify pathogenic agents in the epizootics hitting developing countries. Moreover, sequencing the products resulting from the application of PCR technique on samples of sick animals makes it possible presently to identify the pathogenic agents more precisely.

Veterinary art has acquired an extended palette of vaccines too numerous to be listed here. Among the latest developments in research one must consider: the newest generation of vaccines produced from purified antigens, or recombinant vaccines, work on the genetic resistance to disease, tools for sanitary diagnostics. In the countries of the south priorities are epidemic surveillance and epidemiology in the international network. Unfortunately, poverty and deficiency of technical and veterinary expertise still often lead to resurgence of epidemics or plagues of pests, such as recent cricket invasion of West Africa. This is particularly devastating to underdeveloped countries where consecutive losses (due to parasitic diseases for instance) continue to affect the resources created by animal farming—losses as high as 20% of the livestock.

● *Livestock and environment*
    As we have seen, animal farming, since the Neolithic era, has played, and will continue to play, a preponderant role in human subsistence. As the trend toward greater production and distribution increases, the consequences on the environment, sometimes quite negative, become more important. They are of various kinds.

Let us recall, first of all, that grazing land occupies, according to recent estimates, more than a quarter of the world's land surface and that production of fodder extends over more than one third of arable lands. This extension, which will undoubtedly increase, often has as corollary an intensive deforestation and a certain degradation of the soil. Moreover, raising livestock has a non-negligible influence on climate warming. According to the FAO, it is responsible for 18% of the greenhouse gases, 9% of CO_2 emissions, and above all 37% of methane (CH_4) emissions resulting from the accumulation of porcine excrement and the intestinal fermentation of ruminants. Additionally, the responsibility of 64% of nitrous oxide (N_2O) emissions, as well 64% of ammonia, which contribute to acid rain, is also attributed to livestock farming. Such production also consumes great quantities of water (for growing forage), and causes a variety of pollution (animal waste, fertilizer, pesticides).

Various measures would be appropriate for reducing the environmental impact of livestock farming and its associated activities (grazing, forage cultivation, etc.). Essentially this would involve the restoration of damaged soils through better management of grazing lands, a reduction of the greenhouse gases that arise from deforestation and degradation of grazing grounds, and better management of animal waste ,particularly in industry, in order to avoid water pollution. Better protection of biodiversity (animal) in unexploited zones is also called for.

● *Association between Agriculture and Animal Farming*
There is growing interest (see the IAC[6] report on agriculture in France), on practices associating agriculture and livestock production. The goal is to integrate all of the middle elements (soil, water, nutrients, animal and plant biodiversity) in order to renew resources, limit the effects on the environment and obtain products in sufficient quantity and quality; and all of this to be accomplished within acceptable costs. These objectives correspond to the principles that have been recently termed «the doubly green Revolution». In other words, the issue is to «reconcile the various objectives of preserving the environment, of productivity and of coherence with local social dynamics» (cf. RST report of the French Academy of Science, n° 21, «Sciences et pays en développement – L'Afrique sub-saharienne», p. 138, EDP Sciences (2006)).

In Africa, particularly, where it is often necessary to battle against malnutrition and improve agricultural productivity, the international scientific committee of the IAC, after numerous studies and local surveys, suggested several systems to integrate agriculture and livestock, which could respect these criteria. Thus, different recommendations, depending on the region in question:

– a system that integrates the cultivation of maize in association with that of cotton and of livestock;
– a system of cereals/root plants (maize, sorghum, millet, cassava, legumes), associated with animal farming;

---

6. IAC : Inter-Academy Panel Council : a international instance which represents the strategic and decision-making arm of the IAP (Inter-Academy Panel) an umbrella organization for more than 80 Science Academies.

– the association of an irrigation system for the cultivation of dominant cultures such as rice and cotton with market gardening and animal farming (cattle, poultry).

In this chapter we draw the conclusion that many of today's systems of animal production in Africa no longer seem «sustainable». Besides selection and animal health, research should focus on the integration of livestock and plant farming where it could be possible with the use of agriculture by-products. As for the seasonal movement of livestock, research concerning transhumance concerns for the most part the restoration and improvement of pastures, with a greater role being played by social sciences (georaphy, anthropology, economics...)[7].

---

7. All of these being activists in the cause of a better coordination of agronomic research, particularly in Africa. In this sense, the New Partnership for African Development (NEPAD), through its agricultural component the Comprehensive Africa Agriculture Development Program (CAADP), in association with the FAO, the Forum Africain de Recherche Agronomique (FARA), the World Bank, the African Development Bank and the Consultation Group for International Agricultural Research (GCIAR) should be able to contribute to instituting new husbandry techniques that are better adapted to the environment and local needs.

II.3. ENVIRONMENT – ENERGY – BIODIVERSITY

II.3.1. Energy challenges – Greenhouse effects – Renewable energies – Biofuels

II.3.1.1. Energy challenges – Climate change

In February 2007, the IPCC (*Intergovernmental Panel on Climate Change*) presented its general conclusions before the Académie des Sciences in Paris. This document confirmed that our planet had moved into a period of climate change, the intensity of which is likely to increase, and emphasized the role played by greenhouse gas emissions (GGE) generated by fossil fuels over the past thirty years.

Nobody questions the steep rise in average global temperatures. The real question is: "to what extent is this rise due to human activity, natural factors or a combination of both parameters?" That non anthropogenic factors are contributing to global warming is <u>suggested</u> by observations concerning the melting of quaternary glaciers. The cause was imputed to fluctuations in solar activity, a "natural" effect indeed and, more recently, the combined effect of volcanic activity (and, it's true, aerosols!). However, we are not in a position to say whether these "natural" causes are solely responsible for the variations observed up to 1975.

Since 1975, we have witnessed a sudden change of climate. Specialists are seeing evidence (or consequences) in the faster rise in mean global temperatures observed today, in the occurrence of a series of sudden geochemical events like heat waves, drought or floods in countries previously known to be temperate, and in the increasing frequency of storms and hurricanes, the shrinking of many mountain glaciers, the reduction in volume of pack-ice and polar ice-caps. Most, if not all, specialists find these phenomena difficult to explain without reference to a greenhouse effect linked to human activity and particularly the accelerated production of GGs, resulting from increased consumption of fossil fuels. Indeed, since the mid-1970s, annual GG emissions have increased by 50% (and slightly more for CO_2). It is estimated that CO_2 production has reached about 30 Gt (gigatonnes!). Other important sources of GGE are in agriculture: these are methane and nitrogen oxides resulting from massive deforestation (which also, naturally, indirectly causes an increase in CO_2 levels).

It is estimated that by 2050 annual CO_2 emissions will be between 50 and 60 Gt. This should result, on the one hand, from the fact that certain large industrialised countries intend to meet the increased demand for energy over the next few decades by using traditional coal-fired power plans and, on the other hand, from the fact that transport (of people and goods) will consume more oil-based products and will do so for a long time to come. According to the United States department of energy (DOE), electricity consumption will more than double by 2030 (with similar consumption in countries outside the OECD).

Global annual energy consumption is today estimated at 10 billion tonnes of "oil equivalents" (or 10 giga TOE). It should reach as much as 15 GTOE in 2030 and 22 GTOE in 2050, if no steps are taken on a global scale, particularly since the main emerging countries (China, India, Brazil) are in the process of major economic growth and are also the most populous countries in the world (they are not signatories to the Kyoto protocol and will probably use their large coal reserves).

The measures envisaged, particularly by countries which signed the Kyoto protocol, include: savings in the consumption of energy generated by fossil fuels, capturing and storing CO_2, as well as diversifying energy sources, awarding more space to renewable energies which are not GG emitters or only very slightly. Before discussing this question even briefly, you must be aware that *"no matter what we do, it is already too late to prevent the early climate changes which we are already observing and which are expected to develop by 2020"* (in *"Énergie 2007-2050. Les choix et les pièges"* – Académie des Sciences, Bernard Tissot et coll., tome X, 2007).

As for energy saving, the International Energy Agency (IEA) recently estimated that global savings achievable by 2050 could be between 15 and 35% of current global consumption.

In this respect, major energy savings can be made in the industrial and tertiary sectors which, in the "rich" countries, represent up to 60% of electricity consumption. These savings (which we shall not dwell on, given the general perspective of this book) concern, above all, premises heating, building construction, heat insulation and land-based transport.

The experts feel that CO_2 capture and storage is the only way of ensuring a possible and <u>sustainable</u> substitution of oil-based products to coal. It would allow sustainable use (by reducing the deleterious effects of this industry on the environment). However, as has been seen, electricity production, at least the ¾ which are produced in countries outside the OECD, will probably be derived from coal, although nuclear power will also contribute. Considering that global emissions of CO_2 generated by the use of fossil fuels exceeds 25 billion tonnes and should reach 50 billion by 2050, it has been calculated that <u>10 to 20 billion tonnes of CO_2 must be captured per year and stored, and this for several centuries</u>! (Tests currently under way in the North Sea and Canada concern only 1 to 2 billion per year). In addition the problems caused by injecting CO_2 into deep geological layers (rock porosity, geochemical reactivity, pH effects, water-tightness) are far from being solved. Injecting CO_2 into the deep ocean may have extremely negative effects on the marine ecosystem.

In any case, it is estimated that CO_2 storage can only be envisageable around 2030.

II.3.1.2. Non-CO_2 emitting energy sources

We are now turning towards energy sources likely to generate electricity <u>without carbon dioxide emissions</u>. These are essentially of two types: <u>nuclear energy</u> and <u>renewable energies</u>, to which can be added, though in the fairly distant future, this new vector of storable, clean energy – <u>hydrogen</u>, which is planned for use in transport (IC engine or fuel cell). Nuclear power can supply considerable quantities of electricity. Thermal neutron reactors have a 30 year long period of use, about to be extended to 40 or even 60 years, and there is about a century's worth of Uranium 235 reserves. New generations of Uranium 238 reactors, fast neutron reactors, are expected during the next 50 years. But the problems raised in accepting this type of energy are still present. In short, they require a very heavy investment compared with that of a thermal power plant, disposal of radioactive waste, the risk of proliferation of nuclear weapons and the opposition shown by environmental defence organisations. Germany and the Scandinavian countries plan to close existing reactors permanently over a period of 30 years, but the serious ecological risks evoked by the use of conventional thermal power plants (coal, natural gas) could lead to a revision of this timetable (B. Tissot et coll. mentioned above).

II.3.1.3. Renewable energies

Apart from sources of electricity which do not produce GGs, such as "hydroelectricity" and "nuclear power", thoughts are focusing with increasing frequency on power produced by renewable energy sources. These include solar and wind power which are evoked most commonly to meet a "concentrated" and increasing demand for electricity. The use of these energy forms suffers, however, from their intermittent nature (wind not strong enough in the case of wind power, wind and clouds for the other) so that using these sources means future development of processes for storing electricity, if not on a large scale, as we do for power produced at night by nuclear power plants, at least on a moderate scale (which has not yet been fully developed).

• *Photovoltaic*

However, concerning the use of photovoltaic energy, although installation and maintenance costs are prohibitive for developing countries, we must acknowledge that they could be used to supply small quantities of electricity in areas not connected to the grid, which would allow, for example, a medical branch to be installed for cooling purposes, to preserve reagents and notable vaccines, etc. (see COPED-UNESCO symposium "Énergie solaire et santé dans les pays en développement" ("Solar energy and health in developing countries"), Académie des sciences, Pub. Editions Tec and Doc Lavoisier, 1998). Various countries are beginning to increase the use of solar panels considerably.

• *Biomass*

Biomass is already a non-negligible factor through its traditional use as wood for heating. It covers 10% of the world's energy requirements. This is a form of "direct" use which could continue, particularly in poorer countries with forest reserves, but it also poses ecological problems (deforestation).

Furthermore, for more than fifty years, major programmes have been developed to convert cultivated biomass into biofuels. This involves, for example, so-called "first-generation" fuels produced from oleaginous crops (rape, sunflower), the seeds of which provide oil which is directly usable in farm vehicles or after transformation (diesel engines). It is known that ethanol can also be obtained by fermenting cereals or sugar cane (Brazil). However, apart from the latter example which gives a satisfactory energy balance, serious reservations have been made by various experts concerning global yield linked to energy produced by other forms of cultivated biomass. Furthermore, in the context of the current food crisis, other serious reservations are starting to be made.

- *"Fuel versus food"*

    Indeed, a global risk seems to be developing in this area: this involves growing competition between production for self-sufficiency and fuel supplies (fuel versus food) and opinion is moving increasingly towards arbitration concerning the use of agricultural resources for food, energy and industry. The problem should become even more acute in that the areas suitable for growing "fuel" crops in view of the increasing demand for cars and by industry are very small on a planetary scale. They only occupy about 10% of the 1.5 billion hectares cultivated today across the world. This leads to the danger of having "energy" production move into the area used for food crops which is already inadequate (see II.2.1.2.).

    According to a recent article in The Guardian (29 August 2007), the conclusions of which were taken up by Le Monde on the same day, this threat seems to be very serious.

    Indeed, the United States, the biggest maize exporter in the world, envisages reaching a production equivalent to 35 billion gallons of non-fossil fuel by 2017 to cover some of its transport requirement and reduce its dependency on oil imports. It is estimated that in 2006, 20% of the total US maize crop was converted into ethanol (for only 2% of drivers' needs!). The price of maize exported by the United States has doubled in barely 10 months and the price of wheat has increased by nearly 50%. A growing tendency to convert a large proportion of food plants into biofuels is also being demonstrated in large transition countries (China, India), but also in Europe, Japan and South Africa. Food prices have already increased significantly. Still, according to the above-mentioned sources, the price of food in India has risen by 11% in one year. In South Africa, Mexico and China, even bigger increases have been noted, particularly in animal feed which depends directly on agriculture. However, we must remember that the number of people suffering from malnutrition totals 850 million on a global scale (see the chapter "Feed the planet").

    To summarise, the use of food crops such as wheat, maize, sugar beet, rape and even sugar cane for "energy" needs involves a non-negligible risk of creating competition with their traditional role in the food chain. Another type of competition is indicated between the use of water for food crops and for biofuel production. Only the conversion of sugar cane records a high level of energy efficiency (ratio between the energy produced as biofuel and the energy expended in producing it: plant production and transformation), this being between 6.5 and 7.5. These values are much lower with other crops (around 1.3 to 1.7).

• *Second-generation biofuels*
    Researchers have therefore worked to produce 2nd-generation biofuels on a large enough scale, biofuels which are likely to be better accepted by society as a whole than those produced by the transformation of food crops.

    Indeed, the use of all the lignocellulose biomass could privilege land which is unsuitable for food production. Agricultural or forestry waste could also be used, such as: straw, sawdust. The envisaged conversions generally lead to glucose, obtained by enzyme hydrolysis and transformed into ethanol by fermentation (yeasts).

    Another way is to gasify the lignocellulose biomass ($CO+H_2$) followed by chemical synthesis leading to various fuels.

    It has also been envisaged to use perennial plants such as fescue or Provence cane. They do not need irrigation and have a beneficial effect on soil quality.

    Although the conversion of lignocellulose is doubtless a promising method for providing future biofuel, though this can only be partial, new research (e.g.: genetic, intended to modify the composition of lignocellulose products) and technical progress will be necesary to develop sufficiently economical transformations.

II.3.2. BIODIVERSITY

II.3.2.1. Identifying and protecting biodiversity

II.3.2.1.1. General data – Threats and concerns for a common heritage

    Of all the major crises which threaten the environment today, in terms of general acceptance, the threat to Biodiversity is one of the most severe, and is, in any case, the one whose the effects on sustainable development are already the most marked.

    The word "Biodiversity" is recent in application. It was introduced during the 1960s, by the great American systematician E. O. Wilson. It is absolutely essential to take care of biodiversity; at a first sight the need for this may not immediately seem to be obvious, but the world population is beginning to become aware of it. There are many reasons. We all know that terrestrial and marine ecosystems provide many resources (such as grains, textile fibres, etc.). The fertility of agricultural land and CO_2 absorption depends on their functioning correctly and their stability is vitally important. Our daily news illustrate how disturbing this system stability can cause a multitude of problems such as: floods, tsunamis, tidal waves, reduced resistance to emerging diseases, biological invasion of all kinds or the action of destructive crop pests.

Focusing only on the main aspects of the issue, two global geographic ecosystems deserve particular attention: the world's forest ecosystem, the biggest being the Amazon, and the oceanic ecosystem. Indeed, the Amazon is one of the most heavily forested regions in the world. According to recent studies, 1,200 species of trees have been identified in Guiana, compared with less than 100 in Western Europe. The mass of carbon stored in these trees is enormous, close to 10 gigatonnes. The Amazon forest which, with other tropical forests, hosts half the world's living species, also plays a major role as climate regulator. The threats hanging over this "planetary lung" are well-known: they are due to excessive forestry operations, mining activities, dam construction and sometimes large-scale land clearance for agricultural use. The Brazilian government is making huge efforts to protect this immense forest resource (1/3 of the Amazon region is protected).

The oceans are another global ecosystem, rich in a wide variety of species (about 200,000), but particularly threatened by human activities. Ocean biodiversity is suffering from the effects of pollution, increasing maritime traffic (close to a factor of 5 over 50 years) and over-fishing. The oceans are too often used as receptacles for human waste and are suffering the effects of eutrophication in estuaries (nitrates, phosphates). Another factor in biological deterioration, and not the least important, is the increase in human population in coastal areas! It is estimated that by 2025, 75% of humans will live on a 100 km wide coastal strip, with global population increasing by 15% in less than 20 years. The consequences are often disastrous. Thus, 50 to 90% of large predatory fish (including tuna) have disappeared, gradually dominated by smaller and smaller specimens.

• *The effects of global warming*
But other dangers already threaten, or will soon threaten, Biodiversity as a whole. Everyone knows they are linked to global warming, the origin of which is largely due to or associated with human activities, according to Inter-

governmental Panel on Climate Change (IPCC) experts, who held a meeting a few months ago. The public is familiar with alarms about "greenhouse effects", which can be attributed to increased consumption of coal and hydrocarbons. The economical emergence of countries with high population densities (China, India, etc.), although a welcome progression, must raise general awareness of energy consumption problems. Some models predict that towards the end of the 21st century, global warming will be between 1.5 and 5.8 °C and we know that this is already being felt in the polar regions.

The effects of warming on the geographical distribution of <u>insects</u> are already obvious, with the migration of lepidoptera, coleoptera and dragon flies towards cooler areas.

As for <u>birds</u>, we have accurate data, according to which nearly half the 435 European species listed have migrated North (including the famous white stork).

This movement towards more temperate geographic zones than their traditional habitats also concerns <u>fish</u> (burbot, cod), whereas other species (ray, plaice) are moving to deeper waters. The same observations have been made for geographical movement to more temperate regions by other species (sardines, anchovies...). Similar disturbances linked to geoclimatic effects are expected with <u>plants</u>. Similarly, but unfortunately linked to human intervention, there are many cases of land pollution (Narrow-leaved ragwort) or marine (macroalgae) pollution.

• *Urbanisation, deforestation, extensive agriculture*
Finally, as a general rule, human land management can often become a negative factor for natural biological habitats if it is not reasoned and reasonable. It is enough to think of the <u>considerable expansion of cities,</u> particularly <u>megalopolises,</u> of the construction of major communications networks or, on a different level, of the uncontrolled forest exploitation, that is to say an exploitation without balanced management of tree species and forest renewal. (Every year, 3.4 billion cubic metres of wood is removed); not to mention the tremendous problem of drinking water. Millions of people throughout the world do not have access to it. At the same time, farming, although essential, is a gulf which absorbs nearly 70% of the world's consumption of water, this vital resource so important for biodiversity. If we add up all the factors threatening Biodiversity as the IUCN (International Union for the Conservation of Nature) has done, more than 3,600 plant species and 3, 500 vertebrate species (including ¼ of all mammal species) are threatened with extinction!

• *International attitudes and measures taken*

The international measures taken to protect Biodiversity are held by many specialists to be inadequate. They are none the less wide in scope. Europe has undertaken to reduce the loss of its biodiversity by half by 2010. Many nature parks are being developed (nearly fifty in France). The programme launched by UNESCO thirty years ago: "humanity and the biosphere", has led to the creation of nearly 500 protected sites in a hundred countries.

Many international conferences on the general theme of "Environment and sustainable development" with a major biological component have been held since the 1992 Rio Summit, with commitments to "Millennium Development Goals". These have also largely taken the biodiversity problem into account.

In 2005, the scientific community launched an appeal to set up an international commission of experts known as IMOSEB (International Mechanism of Scientific Expertise on Biodiversity) which was supported by President Chirac at the UNESCO international biodiversity conference. In October of the same year, another meeting on the same theme was held in Mexico, organised by the interdisciplinary committee DIVERSITAS, affiliated to the ICSU (International Council for Science).

II.3.2.1.2. The variety of living species – An unfinished investigation

What exactly do we know about Biodiversity? What are the scientific data available on the number and variety of living species? How does modern Biology learn about their evolutionary relationships and what can be done, today, to improve the conservation of species, continuing their inventory and protection? Do molecular biology and genomics have a role to play in this affair? This is what we now propose to examine.

Science owes a lot to taxonomists and systematicians for their work on the diversity of living species since ancient times. This real quest for the explosion of life began towards the 4th century BC with the observations and initial classifications made by Aristotle and his pupil Theophrastus (see introduction). As Science progressed, and with it the means of communication, the explorations in distant countries, the creation of botanical gardens, collections of rare specimens, Natural History museums, etc., the number of living species listed and classified increased constantly (see the RST report of the Académie des Science on Systematics 2005).

Referring to the world of eukaryotes (we shall see that the world of microbes is undergoing a particular renewal), the total number of species "described and named" at present is thought to be around 1.7 million. These include about 270,000 species of higher plants, 150,000 species, of less evolved plant families: algae, fungi, and more than a million animal species including arthropods (which includes the insect kingdom), by far the most numerous, and vertebrates, with only 45,000 species. However impressive these numbers may seem, this inventory representing decades or centuries of intensive work, two important observations can be made:

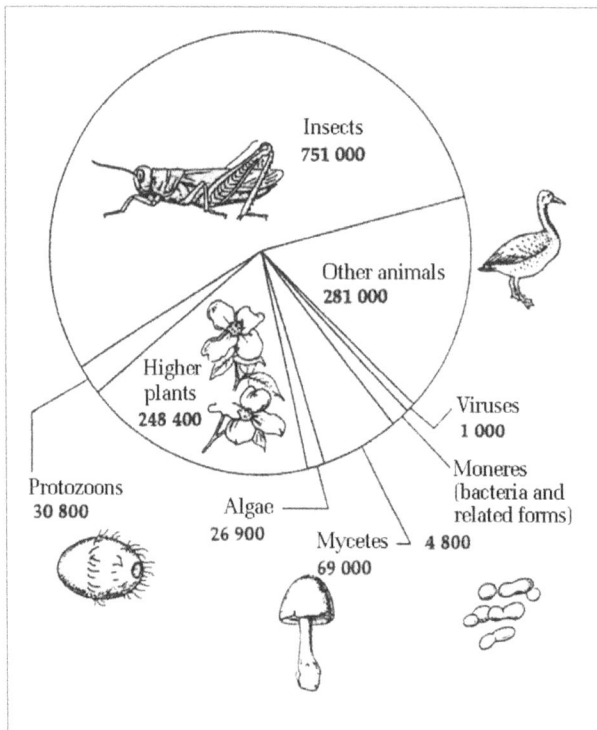

Fig. 8. *Numbers of living species corresponding to various types of organisms currently identified (2002 data)*

These data are taken from an article by Michel Baguette, published in Probio-revue, n° 4 (2001) and duplicated in "*Biologie-Géologie*" APBG, a quarterly educational journal n° 3, p. 579 (2002). The numerical data are provided by E. O. Wilson (1992). At present, the total number of known species is close to 1700,000 (see the text).

Firstly, the total of 1.7 million is considered by all taxonomists to be heavily under-estimated! Indeed, it is assumed, as Simon Tillier, director of the "Species and Speciation" team at the national Natural History Museum writes: "that there are between 6 and perhaps 30 million species, which means that 70 to 95% of life has not yet been listed". There is, indeed, a huge biodiversity which has not yet been described, which is relegated to ecosystems are particularly difficult to access (e.g.: organisms living deep underground or in marine environments, plant and animal parasites, etc.). Every year, about 15 to 17,000 new species are described. This is both considerable and very limited if one takes into account the inventory which still has to be drawn up. In this respect, we must emphasise that establishing the existence of a new species with certainty is not an easy task. The risk of rediscovering "synonymous" species is significant. For example, it is estimated that describing a new species of phytoplankton about 0.5 to 180 micrometers in size occupies one person working full-time for a year! Therefore, great efforts are made internationally to set up networks for data communication and exchange of specimens, a huge task in which information technology now plays a major role.

II.3.2.1.3. Phylogenetic relations – Genomic comparisons

The second general remark, which is fairly obvious, is that "describing is not knowing". Indeed, knowing a species involves having information on its biology, geographic distribution, way of life, its role in the ecosystem and its benefits or application relative to our own species. An important aspect in studying biodiversity lies in what specialists call phylogeny or phylogenesis, a scientific term invented in 1894 by Haeckel, in reference to the relationships between species which are established during evolution and have given rise to the existence of taxons. Drafting phylogenetic trees (dendrograms) is very interesting both to facilitate the classification of newly described species and establish whether or not they have the same ancestry (monophyletic group), but also to support or refute certain hypotheses, learn about the mechanisms of evolution, etc.

To reach this objective, scientists have, for a long time, used morphological or comparative anatomical criteria or degrees of compatibility in crossbreeding, even comparing reproductive systems. But it is precisely at this level of research that molecular biology, classical genetics and, more recently genomics have made some very impressive contributions.

Molecular techniques and databases now provide a way of examining and quantifying biodiversity through the acquisition, storage and comparison of linear sequences of amino acids and nucleotides from different species. The deductions to be made by comparing these sequences does not only lead to the definition of "dichotomies" characteristic of "evolutionary trees". They also provide quantitative measurement of the degree of phylogenetic parentage and consequently, information on the "antiquity" of individual genetic lineages.

The first great successes obtained using the molecular approach to systematics were due to the pioneer work of the American biologist Carl Woese (1987) in the field of microbes. They used comparisons between RNA sequences from the small ribosome subunit (rRNA), 16S RNA. This RNA is relatively "preserved", i.e. its global sequence has undergone few changes during the course of evolution, even over long periods of time, while containing a large number of variation sites (at more than 2000 positions) which facilitates phylogenetic characterisation.

• *The Archaea – The appearance of eukaryotes*
      C. Woese's studies particularly revolutionised our ideas on the early events of evolution. Indeed, they showed that, contrary to the standard distinction involving the existence of 5 main biological kingdoms: plants, animals, fungi, protists and bacteria (Whittaker, 1969), molecular studies now consider three categories or primary ancestral groups: eukaryotes, bacteria and Archaea (Woese, 1990). The Archaea are forms of unicellular life, intermediate between classic bacteria and eukaryotes. In particular, their ribosomal RNA is close to that of eukaryotes and their DNA sometimes includes introns. The Archaea (previously known as "Archaeabacteria") particularly include "extremophile" bacteria, capable of living under extreme conditions of temperature, salinity, etc.

The comparative study of genes coding for 16S rRNA led to some unexpected conclusions. For example, instead of being of recent evolution, as had previously been thought, eukaryotes actually represent a lineage which is doubtless as old as that of the Archaea and Eubacteria. Older lineages are represented by organisms without mitochondria, golgi apparatus and complex cytoskeletal networks (diplomonads, microsporidia, trichomonads). They were followed by a series of independent branches of protists, then the kingdoms of fungi, plants and animals, followed as well as by two particular "phylogenetic groups": stramenopila including red and then green algae (similar to plants through their chlorophyll) and alveolates (ciliates, dinoflagellates) including photosynthetic and non-photosynthetic forms. The almost simultaneous evolutionary separation of this "plants, animals, stramenopila and alveolates" group (sometimes called "eukaryotic crown") from older branches, is thought to have occurred about a billion years ago and it is obvious that, in this assembly of crown eukaryotes, it is the plants and animals which share the most recent history.

Many other inferences relative to evolutionary history have been made possible by the comparative study of sequences of genes coding for ribosomal RNA. They are too numerous and too specific to be mentioned here (see M. L.

Sogin and G. Hinkle, in Biodiversity II, chapter 8, p. 109, 1997 edited by M. L. Reaka Kudla, D. E. Wilson, and E. O. Wilson).

The knowledge that led to partial or complete genome sequencing has also enriched "systematics", a discipline which had been somewhat neglected. For instance, biologists used rapid gene discovery techniques to identify transcribed genes. This consists of obtaining partial sequences from cloned copies of "cDNA" taken from a vast sample store of "messenger" RNA, chosen at random. These partial sequences, generally including 200 to 400 base pairs, sort of molecular labels, are known as ESTs (expressed sequence tags). Details of these ESTs can be obtained through automatic sequencing and thus provide a provisional identity of genes expressed in cells, tissues or complete organisms. These identities are then confirmed by consulting databases containing previously described DNA sequences and proteins obtained from various laboratories, where they are stored (Genbank, EMBL, Swiss Prot, etc.). The details obtained from gene samples expressed via their ESTs lead to useful information for making comparisons relative to phylogenetic relations (see C. J. Bult *et al.*; C. Fields and J. C. Venter, in Biodiversity II, chapter 20, p. 289, 1997).

Although EST sequencing programmes have led to the discovery of many genes, the corresponding databases do not contain enough information about genes which are not often transcribed! The only way of accessing all genes is still to determine the complete genome sequence.

II.3.2.1.4. Plant Genomics and Biodiversity

Great progress has been made with modern approaches to plant diversity, for various reasons linked to both purely scientific and also horticultural or agrifood considerations. Remember that in plants using chlorophyll there are three types of genome: nuclear, mitochondrial and chloroplastic. The genetic organisation and content of the chloroplast genome varies very little from one species to another with their size ranging between 150 and 300,000 base pairs. Chloroplast DNA in tobacco, rice, maize and pine has been fully sequenced.

On the other hand, there are large size variations in plant mitochondrial genomes (from 200 to 1,000 kilobases). Plant nuclear genomes, on the other hand, reflect great plasticity. Plants are usually "diploid" like animals (the somatic cells include one male and one female copy), but their ploidy (number of copies of chromosomes) may vary considerably. For example, we know that many plants domesticated for agricultural or horticultural use are often multiploidal. The size of their genomes varies widely ; e.g.: *Arabidopsis thaliana*, a dicotyledon in the *Brassicaceae* family, which is the first plant whose genome has been fully sequenced (in 2000), only has the equivalent of 140 megabases (millions of base pairs) in its nucleus and about 25,000 genes. The nuclear genome of rice (*Oriza sativa*) has 430 megabases (and about 50,000 genes), that of maize (*Zea maics*) has 2 500 megabases (2.5 billion base pairs), barley has 5,000 megabases and the diploid tulip includes in its heritage 20,000 megabases (20 billion base pairs)!

However, these differences are often due to the existence of many mobile elements (transposons) inserted between genes, as well as to frequent duplication of chromosome or gene segments, which leads to the formation of groups of genes at a single locus. Another very interesting phenomenon involving evolutionary mechanisms, particularly in plants, is what is known as **synteny**.

Indeed, in the early 1990s, it was observed, particularly in cereal plants, that the order of genes along the chromosomes in different species could be preserved (the "synteny" of the genes is preserved). This led to a model of the genomic organisation of cereals in which the various species' chromosomes are distributed in concentric circles and aligned with each other according to their homologies. For example, this model can predict the equivalence between a segment of wheat or maize chromosome and rice or sorghum chromosomes. But this chromosomal transposition of groups of aligned genes, with similar functions from one species to another, was then observed in dicotyledons: therefore, each chromosome of *Brassica oleacea* (cabbage) or *Brassica napus* (Swede and turnip) can be represented by homologous segments on each *Arabidopsis* chromosome! Syntenic segments are also found between *Arabidopsis* and beetroot, potato, sunflower and almond. These preserved segments, transposed from one species to the next are probably traces of the ancestral dicotyledon genome. Thus, the concept of synteny is not only very interesting for understanding the evolution of plant genomes; it can also be used to easily find an interesting gene in a cultivated species relative to the model species (M. Delseny, *Curr. Opin. Plant Biol.*, 7, 126, 2004). The genomes of many plant species are therefore starting to be known completely or partially. Apart from the model plant, *Arabidopsis* and major cereals such as rice and maize, we can mention: barley, millet, sorghum and wheat as well as alfalfa, cotton or soy, tomato and trees such as apple, poplar, pine, fir. It was noted that in 2005, structural genomic studies concerned the equivalent of 60 billion base pairs in plants, the order of which has been established. A very large number of new genes have been discovered, the study of which has led to the development of processes as varied as crossing, synthesis and protein storage in seeds, resistance to viruses, thermal stress, drought, herbicides...

II.3.2.1.5. Animal Genomics and Biodiversity

In the animal kingdom and humans, genomics has also provided complementary information on the degree of gene preservation throughout evolution and, by using animal models, about the mutations responsible for genetic diseases, genes for susceptibility or resistance to diseases, the regulatory networks involved in tissue development and in cell death, the occurrence of cancers, ageing, etc. The genomes of the following species have been fully sequenced: nematode, drosophila, *Aedes aegyptii* (parasite vector), mouse and the genomes of many domesticated animals (e.g. the pig) are the subjects of advanced study. Finally, remember that the start of the 21st century saw the production of a first version of the human genome as we have already mentioned and described previously. Indeed, knowledge of the human genome opens greater prospects for future medicine. This point was brought up in some chapters concerning genetic diseases, for example, and cancers as well as attempted gene therapy.

II.3.2.1.6. Biodiversity of microorganisms – Metagenomics

However, it is doubtless the kingdom of microorganisms, particularly the bacteria and viruses, which has recently recorded the most spectacular progress in investigation of their biodiversity and corresponding genomic information. First of all, hundreds of bacterial or *Archaean* genomes have been sequenced, providing sometimes very important information about the causes of the virulence demonstrated by some of these microorganisms (e.g.: pathogenic forms of *E. coli*, *Mycobacterium tuberculosis*, *Mycobacterium lepre*) and about new vaccine strategies. Exhaustive knowledge of their genomes is also an essential stage in the fight against viruses responsible for emerging diseases (Ebola, Rift Valley fever, etc.) or very dangerous zoonoses (AIDS, bird 'flu, SARS, various acute respiratory infections). It is often essential to study viral genomes, to establish a proper classification and develop appropriate treatments (AIDS).

However, in recent years, bacterial and viral genomic has been the subject of a veritable methodological <u>revolution</u>. This revolution is one of the most robust and already most fruitful forms in the modern approach to Biodiversity. Indeed, "bacterial diversity is fantastic; much greater than the number of bacterial species identified by cultures on appropriate media would lead us to suppose" (P. Vignais in "Science expérimentale et connaissance du vivant", p. 268, edit. EDP Sciences, 2006).

The explanation of this enormous deficit in our knowledge of the "microbial world" has only recently been clarified: the proportion of bacteria that can be cultured and hence identified, is very small. Some of them grow very slowly or require growth media which are so complex that we don't even know what they are. In other cases they are bacteria living in "commensalism", i.e. they can only grow in complex ecosystems involving other bacteria. Often they are strictly anaerobic bacteria which can only be isolated using complex technical assemblies, or they are "extremophiles", requiring very special conditions of temperature or osmolarity. The result is that we only know a very small percentage of the microbial species which live on our planet, perhaps barely more than 1%!

It is precisely to enter this world, which is largely unknown, that new methodological approaches have been designed (and successfully applied), inspired directly by genomic sequencing techniques and known as "Metagenomics or Environmental genomics".

Generally, a heterogeneous sample of bacteria derived from complex ecosystems, assumed to contain a very large number of species is used; the DNA is extracted from this assembly and amplified using the PCR (Polymerase Chain Reaction) technique. Once amplified, this global DNA is sequenced and the data obtained are processed by computers, comparing them with sequences characteristic of microbial species already identified and stored in the database. Of course, the metagenomic procedure is also applicable to the identification of as yet unknown species of protozoa, viruses or Archaea.

Metagenomics has already recorded a fairly large number of achievements. In 2004, J. Craig Venter and his associates (*Science*, 304, 66, 2004) investigated the microbial biodiversity of the Sargasso Sea, close to Bermuda. A total of one billion non redundant base pairs were characterised, annotated and analysed to elucidate their gene content, diversity and the relative abundance of organisms in the samples taken. These sequences were interpreted as coming from 1 800 genomic species, on the basis of the relations between sequences, including 148 unknown bacterial species. In this work, the authors identified more than 1.2 million previously unknown genes, 782 of which code for Rhodopsin type photoreceptors. Another example: in 2006, E. F. Delong et coll. (*Science*, 311, 496, 2006) investigated plankton biodiversity in North Pacific waters.

But other complex environments have been subjected to investigation by metagenomics with respect to the diversity of microorganisms present in them. For example, various work has been done on soil microbiology (M. R.

Rondon et coll., *Applied and Environmental Microbiology*, 2541, 2000; V. Gewin, *Nature*, 439, 384, 2006). Other studies have been performed on the diversity of RNA viruses in intestinal flora (Tao Zhang et coll., *Plos Biology*, 4, 108, 2006) or on populations of bacteriophages (M. Brettbart et coll.; *J. of Bacteriol.*, 185, 6220, 2003).

The study of metagenomes has sometimes shed light on microorganisms which play a previously unsuspected role in the Nitrogen cycle (e.g. bacteria in the *Anemox* genus or in anaerobic methane oxidation (ANME *Archaea*: S. J. Hallon et coll., *Science*, 305, 1457, 2005). New pathways are thus being traced in this investigation of the major biogeochemical cycles.

There is no doubt that the global systematic approach to learning about complex microbial populations through the use of biochemistry, genomics and bioinformatics will considerably enrich our knowledge of the infinitely small in biology, with unsuspected theoretical and practical consequences.

CONCLUSION

Very often during its history, biology has followed current trends, styles or concerns.

"Friendly science" in the 18th century, biology was often the work of erudite people, philosophers advocating a return to nature, with a predilection for botany, herbaria, large collections of rare specimens. The inventory of plant and animal species living in distant countries became extremely fashionable! The "Jardin Royal" – previously "jardin des herbes médicinales" (garden of medicinal plants) – soon became the great establishment which we know as the Muséum national d'Histoire Naturelle (national Natural History Museum), where scientific life reigned supreme and the great naturalists were gradually drawn to it to make their reputations. Their presence led to in-depth discussions on a scale which quickly outclassed that of the scientific world as soon as the question of the origin of species was raised. Evolutionary theory attracted or revolted, according to their turn of mind, philosophical or religious inclination; its controversial side left no-one indifferent...

But let's take another epoch, the Second Empire in France (1852-1870). At that time, biology developed more industriously as its alliance with chemistry became more obvious. Pasteur was not only the famous adversary of spontaneous generation. All of his work was marked with the stamp of solid application and the desire to meet national interests (before moving on to humanity as a whole). Therefore, he worked on preserving wine without altering its bouquet, improving "French" beer (very inferior to that across the Rhine according to Napoleon III), combating threats to silk worms, then protecting society during these times of puerperal fevers, post-operative septicaemia, zoonoses. In short defining the first rules of social hygiene before reaching the work that would make him famous: the development of vaccinations. All this tinted with a certain opposition to German competition followed by an avowed desire for revenge, after the taking of Sedan! In Pasteur, his duty as a citizen was never far from his scientific work.

As time went by, we reached the start of what was later called the era of modern biotechnology. Industry already wanted to take over Pasteur's theories and methods, and not only in France. A bio-industry of fermentation began to prosper. The remarkable rise of German enzymology accentuated the movement even more. The first patents based on life sciences were filed[1]. Without going so far as to claim that it was identified with an activity at the service of competing national economies, towards the end of the 19th century, biology was presented, through its applications, as the heir of positivism and "St Simonian" industrialism.

A century later, with the birth of molecular biology and the growing power of genetics, then the first achievements of genetic engineering and transgenesis, biology has not only moved to the forefront of modern science, but is now penetrating deep into the social and ethical fields, while enriching medical and agricultural biotechnology. The first debates and fierce reactions in the United States and Europe to early genetic engineering, to the extent that the first moratoria and mandatory confinement measures were applied to bioscientific research, would doubtless be remembered. This soon supplied fuel to a new area of scientific ethics. The word "bioethics" soon became familiar, going sometimes beyond, in virtue of its resonance, its adequacy to society's aspirations and its wide range of application, the strictly medical deontology, which was strongly attacked at the end of the Second World War at the Nuremberg trials. On the one hand, through its promises revealing the benefits of medicine, agriculture and industry, but on the other through the reservation or opposition it generated, new biology, inspired by molecular biology and genetics, is now attracting a public which had, for a long time, remained indifferent to its advances. It is also clear that most important State institutions, everywhere in the world, have taken this field over to some extent! The law has never before paid so much interest in biology nor has legislation promulgated so many laws to set limits to its application. It is not only the Summit meetings held between major industrialised powers which regularly discuss the social acceptability of bioscience – and the new problems it raises. Henceforth the life sciences – and the situation is even more marked with stem cell or transgenic plant research, for different reasons it is true – will be at "the heart" of the news and the media. An unmistakeable sign in these times where communication holds pride of place...

---

1. The first by Pasteur himself...

But – referring at last to what this very book is about – what stirs society, nations, industry and general opinion today is, above all, everything concerning the major problems affecting the whole planet. This is also, consequently, what is encapsulated in the incantatory but nevertheless inventive concept of "sustainable development". Here, it's true, biology is not alone at the forefront! All scientific disciplines and their supporting techniques are now involved. In its political, economic acceptance and its socio-ethical dimension, the world is turning to science to find fresh support. How can we take on this famous development which commits humanity for several generations? What solutions can be found to cope with natural events and those resulting from the often uncontrolled actions of humanity as well as wars, terrorism and the overpopulation of towns, the lack of foresight with respect to biological balance, the constantly increasing production of greenhouse gases? Finally, how can we respond to the millennial goals so clearly defined in Johannesburg in what is most significant because it often addresses the most desperate situations: illiteracy, malnutrition, maternal and infantile mortality? But although all the forces of science and human engineering are concerned, the life sciences should be in the lead, because, if biology is not the only science which can analyse and predict risks and propose remedies, it still plays an essential and central role when it involves taking into account certain aspects of those which most clearly demonstrate the planet's acknowledged fragility: those which threaten human and animal health, affect biodiversity, allow inequality to come before access to food resources, or compromise the sustainability and renewal of ecosystems.

We shall not linger over the actual or expected contribution of biology in this new context. The book has often echoed them. It seems to us that biology has rarely been so in tune with its times. Indeed, it is moving into a new phase which should make it more attentive than ever to the hopes and fears of a human community which is called on to face new risks, risks which are all the more serious in that they are often evoked on a planetary scale.

This biology, focusing on the great questions of development, will be assigned to two battles which we feel will gradually become inseparable. One is simply the traditional, but absolutely vital, fight proper to any true science: the general quest for knowledge. It is a battle against the immensity of the unknown, an attempt, in this case, to penetrate the arcanes of the complexity of life. The other battle consists in paying more attention to the great challenges of our days, and therefore closer to the efforts that society must deploy in order to reach a more harmonious evolution of the planet on which it lives, a more reasonable, more just evolution, with a less egoistic vision of the present and giving more thought to the future.

BIBLIOGRAPHY

I – THE FABULOUS DESTINY OF BIOLOGY

I.1. HISTORICAL OVERVIEW

Bernard Cl. : « Introduction à l'étude de la médecine expérimentale », (1865), J.-B. Baillère (préface par F. Dagognet), Garnier Flammarion, (1966)

Buican D. : « La génétique et l'évolution », PUF, Que sais-je ?, (1986)

Descartes R. : « Traité de l'homme », (1664)

Giordan A. : « Histoire de la Biologie », 2 t., Paris, Techniques et documentation/Lavoisier

Gros F. : « L'ingénierie du vivant », Odile Jacob, (1990)

Guyenot E. : « Les sciences de la vie aux XVIIe et XVIIIe siècles : l'idée d'évolution », Paris, Albin Michel, (1941)

Harvey W. : « De motu cordis » (de la circulation du sang) (traduction française par Ch. Richet, 1869), Christian Bourgeois, (1990)

Magendie F., in « Précis élémentaire de Physiologie », (1825)

Mayr E. : « Histoire de la Biologie », 2 t., Fayard, (1989)

Pichot A. : « Histoire de la notion de vie », Gallimard, (1995)

Ratcliff M. : « Le concept de suite d'expériences comme reflet de l'activité naturaliste du XVIIIe siècle », in « Bulletin d'histoire et d'épistémologie des sciences de la vie », 2/1, 11, (1995)

Taton R. : « La Science antique et médiévale. Des origines à 1450 », Quadrige/ Presses Universitaires de France, (1994)

Vignais P. : « Science expérimentale et connaissance du vivant, La méthode et les concepts », EDP-Sciences, (2006)

Vignais P. : « La Biologie, des origines à nos jours. Une histoire des idées et des Hommes », EDP Sciences, (2001)

I.2. MOLECULAR BIOLOGY AND ITS ACHIEVEMENTS

I.2.1. Molecular biology of the gene (double helix, gene expression and regulation, the "Central Dogma")

Avery O.T., McLeod C. and Mc Carthy M. : « Study of the chemical nature of the substance inducing transformation of pneumococcal types », J. Exp. Med., 79, 137, (1944)
Beadle C. and Tatum E.L. : « Genetic control of biochemical reactions in Neurospora », Proc. Nat. Acad. Sci., USA, 27, 499, (1941)
Hunt M. Th. : « Sex limited inheritance in Drosophila », Science, 32, 120, (1910)
Jacob F. and Monod J. : « Genetic mapping of the elements of the lactose region in *E. coli* », Biochem. Biophys., Res. Commun., 18, 693, (1965)
Jacob F. and Monod J. : « Genetic regulatory mechanisms in the synthesis of proteins », 3, 318, (1961)
Kendrew J. : « The thread of life: an introduction to molecular biology », Cambridge, Mass. Harvard, Uni. Press, (1966)
Kornberg A. : « Aspects of DNA replication », Cold Spring Harbor, Symp. Quant. Biol., 43, 1, (1978)
Kuhn Th. : « La structure des révolutions scientifiques » (traduit de l'américain par Laure Meyer, Champs-Flammarion, (1983)
Mendel G. : « Versuchen über Pflanzen-Hybriden », Mémoire présenté devant la société scientifique de Brünn, Verh. Naturforsh. Ver., Brünn, 4, 3, (1866)
Miescher F. : « Uber die chemische zusammensetzung der erberzellen », in Hoppe Seyler's Medicinisch gemische Unterschungen, Berlin, August Hirschwald, 4, 441, (1871)
Monod J., Wyman J. and Changeux J.P. : « On the nature of allosteric transitions: a plausible model », J. Mol. Biol., 12, 88, (1965)
Pauling L., Corey R.B. and Branson H.R. : « The structure of proteins : two hydrogen bonded configurations of the polypeptide chain », Proc. Nat. Acad. Sci., USA, 37, 205, (1951)
Perutz M. : « Proteins and nucleic acids », Elsevier, (1962)
Schrödinger E. : « What is Life ? », Cambridge Press, (1945)
Watson J.-D. and Crick F.H. : « Molecular structure of nucleic acids. A structure for desoxyribose nucleic acid », Nature, 171, 737, (1953)

Periodicals

Debru Cl. : « L'esprit des protéines ; Histoire et philosophie biochimique », Hermann, (1983)
Gros F. : « Les secrets du gène », Odile Jacob, (1986)
Jacob F. : « La logique du vivant une histoire de l'hérédité », Gallimard, Bibliothèque des Sciences humaines, (1970)

Lwoff A. : « Jeux et combats », Librairie Arthème Fayard, (1981)

Monod J. : « Le hasard et la nécessité. Essai sur la philosophie naturelle de la biologie moderne », Le Seuil, (1970)

Morange M. : « Histoire de la Biologie moléculaire », La Découverte, (1994)

I.2.2. The genetic code – The transfer of genetic information: transcription and translation

Basilio C., Wahba A.J., Lengyel P., Speyer J.F. and Ochoa S. : « Synthetic polynucleotides and the amino acid code V », Proc Natl Acad Sci USA, 48, 613, (1962)

Brenner S., Jacob F. and Meselson M. : « An unstable intermediate carrying information from genes to ribosomes for protein synthesis », Nature, 190, 576, (1961)

Crick F.H.C., Griffith J.S. and Orgel L.E. : « Codes without commas », Proc. Nat. Acad. Sci., USA, 43, 416, (1957)

Gamow G.A. : « Possible relation between deoxyribonucleic acid and protein structures », Nature, 173, 318, (1954)

Gros F., Hiatt H., Gilbert W., Kurland C.G., Risebrough R.W. and Watson J.-D. : « Unstable ribonucleic acid revealed by pulse labeling in *E. coli* », Nature, 190, 581, (1961)

Grunberg-Manago M. and Ochoa S. : « Enzymatic synthesis and breakdown of polynucleotides : polynucleotide phosphorylase », J. Amer. Chem. Soc., 77, 3165, (1955)

Hall B.D. and Spiegelman S. : « Sequence complementarity of T_2 DNA and T_2 specific RNA », Proc. Nat. Acad. Sci., USA, 47, 137, (1961)

Nierenberg N.W. and Mathaei J.H. : « The dependence of cell free protein synthesis in *E. coli*, upon naturally occurring or synthetic polynucleotides », Proc. Nat. Acad. Sci., USA, 47, 1588, (1961)

Revel M. and Gros F. : « A factor from *E. coli* required for the translation of natural messenger RNA », Biochem. Biophys. Res. Commun., 25, 124, (1966)

Periodical

Kaplan J.C. and Delpech M. : « Biologie moléculaire et médecine », Flammarion, (1989)

I.2.3. Gene regulation – The repressor – The lactose operon

Gilbert W. and Müller-Hill B. : « Isolation of the lac repressor », Proc. Nat. Acad. Sci. USA, 56, 1891, (1966)

Jacob F. and Monod J. : « Genetic mapping of the elements of the lactose region in *E. coli* », Biochem. Biophys. Res. Commun., 18, 693, (1965)

Monod J. and Cohn M. : « La biosynthèse induite des enzymes (adaptation enzymatique) », Adv. Enzymol., 13, 67, (1952)

Schwartz M. : « Sur l'existence chez *E. coli*-K12 d'une régulation commune à la biosynthèse des récepteurs du bactériophage et au métabolisme du maltose », Ann. Inst. Pasteur, 113, 685, (1967)

Temin H. and Baltimore D. : « RNA directed DNA synthesis and ANA tumor viruses », Advances Virus Res., 17, 129, Academic Press, (1972)

I.2.4. The central dogma of molecular biology

Crick F.H.C. : « On protein synthesis », Symp. Soc. Exp. Biol., 12, 548, (1958)

I.3. GENETIC ENGINEERING – BASIC CONSEQUENCES – APPLICATIONS

I.3.1. Genetic engeneering – Discovery – Biology of higher organisms

Berg P., Baltimore D., Brenner S., Roblin R.O. and Singer M.F. : « Asilomar conference on recombinant DNA molecules », Science, 188, 991, (1975)

Cohen S., Chang A.C.Y., Boyer H.W. and Helling B. : « Construction of biologically functional bacterial plasmids in vitro », Proc. Nat. Acad. Sci., USA, 70, 3240, (1973)

Dintzis H.M. and Knopf P.M. : « Informational macromolecules », Academic Press, New York, (1963)

Jackson D.A., Symons R.H. and Berg P. : « Biochemical method for inserting new genetic information into DNA of Simian virus 40 : circular SV40 molecular containing lambda phage genes and the galactose operon of *E. coli* », Proc. Nat. Acad. Sci., USA, 69, 2904, (1972)

I.3.2. Exons-introns

Berget S.M., Moore C. and Sharp P.A. : « Spliced segments at the 5'terminus of adenorivus 2 late mRNA », Proc. Nat. Acad. Sci. USA, 74, 3171, (1977)

Brody E. and Abelson J. : « The spliceosome: yeast premessenger RNA associate with a 40 S complex in a splicing dependant reaction », Science, 228, 963, (1985)

Cech T.R., Zang A.J. and Grabowski P.J. : « In vitro splicing of the ribosomal RNA precursor of tetrabymena. Involvement of a guanine nucleotide in the excision of the intervening sequence », Cell, 27, 487, (1981)

Cech T.R. : « RNA as an enzyme », Sci. Amer., 255, 64, (1986)

Chambon P. : « Structure et expression des gènes eucaryotes en mosaïque I », *in* « Exposés sur la génétique », C.R. Acad. Sci., 291, supplément, p. 21, (1980)

Chow L.T., Gelinos R.E., Broker T.R. and Roberts R.J. : "An amazing arrangement at 5 ' ends of Adenovirus-2 mRNA", Cell, 12, 1 (1997).

Gilbert W. : « Why genes in pieces? », Nature, 271, 501, (1978)

Kourilsky Ph. : « Structure et expression des gènes eucaryotes en mosaïque II », *in* « Exposés sur la génétique », C.R. Acad. Sci., 291, supplément, (1980)

Le Pennec J.-P., Baldacci P., Perrin F., Cami B., Gerlinger P., Krust A., Kourilsky Ph. and Chambon P. : « The ovalbumin split gene: molecular cloning of Eco RI fragments *c* and *d* », Nucleic Acids Res, 12, 4547, (1978)

Sharp P. : « Splicing of messenger RNA precursors », Science, 235, 766, (1987)

I.4. THE COMPLEXITY OF GENETIC MATERIAL IN "EUKARYOTES"

I.4.1. Chromatin compaction – nucleosomes

Bradbury E.N. : « Structure and function of Chromatin », CIBA Foundation, Elsevier, 28, 131, (1975)

Felsenfeld G., Mc Ghee J. : « Structure of the 30 nm chromatin fiber », Cell, 44, 375, (1986)

Klug A., Rhodes D., Smith J., Finch J.T. and Thomas J.G. : « A low resolution structure for the histone core of the nucleosomes », Nature, 287, 509, (1980)

Kornberg R. : « Chromatin structure: a repeating unit of histones and DNA », Science, 184, 868, (1974)

Richmond T.J., Finch J.T. and Klug A. : « Studies of nucleosome structure », Cold Spring Harbor, Symp. Quant. Biol., 47, 493, (1982)

Rouvière-Yaniv J. and Gros F. : « Characterization of a novel low molecular weight DNA binding protein from *Escherichia Coli* », Proc. Nat. Acad. Sci., USA, 72, 3420, (1975)

I.4.2. Epigenetic modifications

Bird A. : « CpG rich islands and the function of DNA methylation », Nature, 321, 209, (1986)

Cedar H. : « DNA methylation and gene activity », Cell, 53, 3, (1988)

Periodical

« Actualités de l'épigénétique », Biofutur, 243, 18, (2004)

I.4.3. Positive regulation – Promoters – Cis-regulatory sequences

Dynan W.S. and Tjian R. : « Isolation of transcription factor that discriminate between different promoters recognized by RNA polymerase II », Cell, 32, 669, (1983)
Dynan W.S. and Tjian R. : « Control of eukaryotic messenger RNA synthetas by sequence specific DNA binding proteins », Nature, 316, 774, (1985)
Elgin S. : « Dnase-1, hypersensitive sites of chromatin », Cell, 27, 413, (1981)
Kingstron R.E. : « Transcription control and differentiation: the HLH family c-myc and C/EBP », Current Opinion in Cell Biologie, 1, 1081, (1989)
Müller M., Gerster T., Schaffner W. : « Enhancer sequences and the regulation of gene transcription », Eur. J. Biochem, 176, 485, (1988)
Schaffner W. : « Eucaryotic transcription – The role of cis and trans-acting elements in initiation », *in* « Eucaryotic transcription current communications in Molecular Biology »,Y. Gluzman ed., Cold Spring Harbor Laboratory, p. 1, (1985)
Serflying E., Jasin M. and Shaffner W. : « TIG. Enhancers and eukaryotic gene transcriptome », (August, 1985)
Shapiro J.A. and Cordell B. : « Eucaryotic mobile and repeated genetic elements », Biol. Cell., 43, 31, (1982)
Singer M. : « SINE$_S$ and LINE$_S$: highly repeated short and long intersperced sequences in mammalian genomes », Cell, 28, 433, (1982)
Wu C. : « Two protein binding sites in chromatin implicated in the activation of heat shock genes », Nature, 309, 229, (1984)
Yaniv M. : « Regulation of eukaryotic gene expression by trans-activating protein and cis-acting DNA elements », Biol. Cell., 50, 203, (1984)

I.4.4. Coding and non-coding DNA

Periodical

« Initial sequencing and analysis of the human genome », Nature, 409, 860-921, (2001)

I.4.5. Repetitive elements

Baltimore D. : « Retroviruses and retro-transposons : the role of reverse transcription in shaping the eukaryotic genome », Cell, 40, 481, (1985)

Britten R.J. and Kohne : « Repeated sequences in DNA », Science, 161, 529, (1968)

Cavalli L., Lederberg J. and Lederberg E.M. : « An infective factor controlling sex compatibility in bacterium Coli », J. Gen. Microb., 8, 89, (1953)

Finnegan D. : « Transposable elements and proviruses », Nature, 292, 800, (1981)

Jacob F. and Wollman E.L. : « Les épisomes, éléments génétiques ajoutés », C.R. Acad. Sci., Paris, 247, 154, (1958)

Korenberg J. and Rykowski M. : « Human genome organization : Alu, lines and the molecular structure of metaphase chromosome bands », Cell, 53, 391, (1988)

Mc Clintock B. : « Chromosome organization and genetic expression », Cold Spring Harbor, Symp. Quant. Biol., 16, 13, (1951)

Mc Clintock B. : « Genetic systems regulating gene expression during development », Dev. Biol. (suppl.), 1, 84, (1967)

I.5. GENOMICS – GENERAL DATA – CONSEQUENCES – APPLICATIONS

I.5.1. Structural and functional genomics

Belasco J.G. and Brawerman G. : « Control of messenger RNA stability », Academic Press, xviii, 517, (1993)

Capecchi M.R. : « Altering the genome by homologous recombination », Science, 244, 1288, (1989)

Chambon P. : « Séquences consensus d'épissage », cité dans Kaplan J.-C. et Delpech M., « Biologie moléculaire et Médecine », Flammarion, 4, 72, (1993)

Changeux J.-P. and Danchin A. : « Selective stabilization of developing synapses as a mechanisms for the specification of neuronal networks », Nature, 264, 705, (1976)

Lander E.S. *et al.* : « Initial sequencing and analysis of the human genome », Nature, 409, 860-921, (2001)

Maxam A. and Gilbert W. : « Sequencing end labeled DNA with base-specific chemical cleavages », Methods in enzymology, Wu, Moldave and Grossman eds., Acad. Press, 65, 499, (1980)

Sanger F., Nickler F. and Coulson A.R. : « DNA sequencing with chain terminating inhibitors », Proc. Nat. Acad. Sci., USA, 74, 5463, (1977)

Scherrer K. and Jost J. : « Gene and genon concept : a conceptual and informative-theoretic analysis of genetic storage and expression in the light of modern molecular biology », Theory Biosci., 126, 65, (2007)

Venter J.C. *et al.* : « The sequence of the human genome », Science, 291, 1304, (2001)

I.5.2. Genetic Polymorphism – SNP

Cavalli-Sforza L.L. : « The human genome diversity project », in IBC proceedings II., UNESCO, (1995)

Cavalli-Sforza L.L., Piazza A., Menozzi P. and Moutain J. : « Reconstruction of human evolution », Proc. Nat. Acad. Sci., USA, 85, 6002, (1998)

Gros F. : « Les SNPs » *in* « Mémoires scientifiques – un demi-siècle de biologie », Odile Jacob, Paris, p. 232, (2003)

Jeffrey A.J., Wilson V. and Thein S.L. : « Hypervariable satellite regions in human DNA », Nature, 314, 67, (1989)

I.5.3. A Biology of molecular networks : transcriptomes – Proteomes

Amouyal P. : « Vers des profils pharmacologiques », Biofutur n° 206, 86, (2000)

Celis J.E. *et al.* : « 2D protein electrophoresis : can it be perfected ? », Curr. Opin. Biotech., n° 10, 16, (1999)

Garin J. : « Analyse protéomique. Exploration cellulaire et annotation », Biofutur, hors série, n° 4, p. 28, (2002)

Kaplan J.-C. and Delpech M. : « Les techniques d'amplification élective in vitro (PCR) », *in* Biologie moléculaire et Médecine, 2e éd., p. 558, (1993)

Kendrew J. : « The thread of life : an introduction to molecular biology », Cambridge, Mass. Harvard, Univ. Press, (1966)

Minard Ph. : « Ingénierie des protéines », Biofutur, n° 288,

Mullis K.B. and Faloona F.A. : « Specific synthesis of DNA in vitro, via a polymerase catalyzed chain reaction », Methods Enzymol., 155, 335, (1987)

Perutz M. : « Proteins and nucleic acids », Elsevier, (1962)

Van't Veer L.J. *et al.* : « Gene expression profiling predicts clinical outcome of breast cancer », Nature, 415, 530, (2002)

I.5.4. What is a gene ? Systems biology

Dawkins R. : « The extend phenotype: the gene as the unit of selection », Freeman, (1982)

Kitano H. : « Foundations of systems biology », MIT Press, Cambridge, MA USA, (2001)

Laforge B. *et al.* : « Progress in biophysics and molecular biology », 89, 93, (2005)

Noble D. : « The music of Life: Biology beyond the genome », Oxford Univ.
    Press, (2006)
Noble D. : « The rise of computational biology », Nature Reviews Molecular
    Cell Biology, 3, 460, (2002)
Novartis Foundation : « The limits of reductionism in biology », Chichester,
    Wiley, (1998)

I.6. A NEW INSPIRATION IN MOLECULAR BIOLOGY – THE
WORD OF RNAS AND THE PHENOMENA OF INTERFERENCE
– THE RETURN OF EPIGENETICS

I.6.1. The word of RNAs

I.6.2. Si-RNA and micro-RNA

Bartel D.P. and Zheng Chen Z. : « Micromanagers of gene expression: the
    potentially widespread influence of metazoan micro RNAs », Nature
    Reviews (Genetics), 5, 306, (2004)
Fire A., Xu S., Montgomery M.K, Kotsas S.A., Driver S.E. and Mello C.C. :
    « Potent and specific genetic interference by double stranded RNA in
    Caenorhabditis elegans », Nature, 391, 806, (1998)
Jogensen R.A. : « Sense cosuppression : past, present and future », edit. By
    G. Hannon, RNA interference, Cold Spring Harbor Press
Lee R.C., Feinbaum R.L. and Ambros V. : « The C. elegans heterochromatic
    gene lin-4 encodes small RNAs with antisense – complementarity to
    lin-14 », Cell, 75, 843, (1993)
Lewis B.P., Bartell D. et al. : « Prediction of Mammalian micro-RNA targets »,
    Cell, 115, 787, (2003)
Murchison E.P. and Hannon G.J. : « miRNAs on the move : miRNA biogen-
    esis and the RNAi machinery », in Curr. Opin. in Cell Biology, 16, 223,
    (2004)
Palatnik J.E. et al. : « Control of leaf morphogenesis by micro RNAs », Nature,
    425, 257, (2003)
Park W., Li J., Song R., Messing J. and Chen X. : « Carpel factory, a Dicer
    homolog and HEN1, a novel protein, act in micro-RNA metabolism in
    Arabidopsis thaliana », Curr. Biol., 12, 1484, (2002)
Plasterk R.H.A. : « RNA silencing : the genome's immune system », Science,
    296, 1263, (2002)
Reinhart et al. : « The 21-nucleotide let-7 RNA regulates developmental timing
    in Caenorhabditis elegans », Nature, 403, 901, (2000)
Sijen T. and Plasterk R.H.A. : « Transposon silencing in the Caenorhabditis
    elegans germ line by natural RNAi », Nature, 426, 310, (2003)

Tang G., Reinhart B.J., Bartel D.P. and Zamore P.D. : « A biochemical framework for RNA silencing in plants », Genes Dev., 17, 49, (2003)

Wichtman B., Ha I. and Ruvkun G. : « Post transcriptional regulation of heterochromatic gene lin-14 by lin-4 mediates temporal pattern formation in C. elegans », Cell, 75, 855, (1993)

Yonath A. : « Ribosomal crystallography : peptide bond formation, chaperone assistance and antibiotic activity », Mol. Cells, 20, 1, (2006)

I.6.3. The return of epigenetics – When heterodoxy becomes a symbol of openness

Avner P. and Heard E. : « x-chromosome inactivation: counting, choice and initiation », Nat. Rev. Genet., 2, 59, (2001)

Cantoni G. : « S-Adenosyl methionine; a new intermediate formed enzymatically from L-methionine and adenosine triphosphate », J. Biol. Chem., 204, 403, (1953)

Franklin S.G. and Zweiler A. : « Non-allelic variants of histones 2a, 2b and 3 in mammals », Nature, 266, 273, (1977)

Holliday R. and Pugh J.E. : « DNA modification mechanisms and gene activity during development », Science, 187, 226, (1975)

Lyon M.F. : « Gene Action in the x-chromosome of the Mouse (Mus musculus L), Nature, 190, 372, (1961)

Morange M. : « 60 ans d'épigénétique », Biofutur, n° 243, p. 18-31, (2004)

Plath K. et al. : « Role of histone H3 lysine 27 methylation in X inactivation », Science, 300, 131, (2003)

Plath K. et al. : « Xist RNA and the mechanism of X chromosome inactivation », Annu. Rev. Genet., 36, 233, (2002)

Riggs A.D. : « X inactivation, differentiation, and DNA methylation », Cytogenet. Cell. Genet., 14, 9, (1975)

Russo V.E.A, Martienssen R.A. and Riggs A.D., in « Epigenetic mechanisms of gene regulation », Cold Spring Harbor, laborat. Press, p. 1, (1996)

Schubert H.L. et al. : « Many paths to methyltransfer : a chronicle of convergence », Trends. Biochem. Sci., 28 (6), 329, (2003)

Silva J. et al. : « Establishment of histone H3 methylation on the inactive X chromosome requires transient recruitment of Eed-Enx1 polycomb groupe complexes », Dev. Cell., 4 (4), 481, (2003)

Waddington C. : « L'épigénotype », Endeavour, 1, 18, (1942)

II – BIOLOGY AND THE GREAT DEVELOPMENTAL CHALLENGES

II.1. HEALTH

II.1.1. Infectious diseases (the revival of microbiology, vaccines, diagnosis and anti-viral therapy, zoonoses, prion diseases)

II.1.1.1. The return of infectious diseases – Diseases of poverty – Neglected tropical diseases

Periodical

F. Gros (coordonnateur), « Sciences et pays en développement – L'Afrique subsaharienne francophone », Rapport RST n° 21 de l'Académie des sciences (2006)

II.1.1.2. MICROBIOLOGY AND ITS REVIVAL

• *General considerations*
« Le péril infectieux. Quelles stratégies de lutte ? », Biofutur, n° 217, (déc. 2001)
Schatz A., Bugie E. and Waksman E.S. : « Streptomycin : A Substance Exhibiting Antibiotic Activity Against Gram-Positive and Gram-Negative Bacteria », Proc. Soc. Exp. Biol. Med., 55, 66, (1944)
Ullmann A. : « Pasteur et Koch : Distinct ways of thinking about infectious diseases », Features, Microb.2, n° 8, 383, (2007)

• *Factors in microbiology revival*
• *Genomics and virulence*
Hacker J. and Kapper J.B. : « The concept of pathogenicity islands », *in* Kapper J. and B. Hacker eds., « Pathogenicity islands and other mobile virulence elements », ASM press, 1, 11, (1999)
Kaiser J. : « Resurrected influenza virus yields secrets of deadly 1918 pandemic », Science, p. 28, (7 oct. 2005)
Laugier N. : « Les gènes du pathogène (*E. coli*, O157 : H7) », Biofutur, (mars 2001)
Perna N.T. *et al.* : « Genome sequence of enterohaemorrhagic Escherichia Coli O157:H7 », Nature, 409, 529, (2001)
Sansonetti P. : « New millenium, new microbiology ? », *in* Med/Sci., 17, n° 67, 687, (2001)
Tumpey T.M. *et al.* : « A Two-Amino Acid Change in the Hemagglutinin of the 1918 Influenza Virus Abolishes Transmission », Science, 315, 655, (2007)

• *Target cells and the penetration mechanisms of pathogenic bacteria*
Cossart P., Pizzaro-Cerda J. and Lecuit M. : « Invasion of mammalian cells by Listeria monocytogenes: functional mimicry to subvert cellular functions », *in* Trends, Cell. Biol., 13, n° 7, 23, (2003)
Sansonetti P.J. : « Microbial pathogenesis, new paths into a new millenium », Trends Microbiol. 8, 196, (2000)
Tran Van Nhieu G., Bourdet Sicard R., Duménil G., Blocker A. and Sansonetti P. : « Bacterial signals and cell responses during Shigella entry into epithelial cells », Cell Microbiol., 2, 187, (2000)
Tran Van Thien G., Cossart P. : « Détournement des fonctions cellulaires clés par les bactéries pathogènes », Méd./Sci., 17, 701, (2001)

• *Susceptibility genes*
Abel L., Sanchez O., Oberti J., Tuc N.V., Van Hoa L., Lap V.D., Skamene E., Lagrange P.H. and Schurr E. : « Susceptibility to leprosy is linked to the human NRAMP1 gene », J. of Infect. Dis., 177, 133, (1998)

• *Environment and reservoirs of pathogenics agents*
Labigne A. : « Connaissance du génome de *Helicobacter pylori* : implications pour la physiopathologie et la thérapeutique », Med/Sci., 17, 712, (2006)
Rhodain F. : « Rapport sur l'évaluation du risque d'apparition et de développement de maladies animales compte tenu d'un éventuel réchauffement climatique » *in* ECRIN, 68, 23, (2007)
Schwartz M. et Rhodain F. : « Des microbes ou des hommes, qui va l'emporter ? », Odile Jacob, (2008)
Tiollais P., Charnay P. and Vyas G.N. : « Biology of hepatitis B virus », Science, 213, 406, (1981)
Tomb J.F., White O., Kerlawage A.R. *et al.* : « The complete genome sequence of the gastric pathogen, Helicobacter pylori », Nature, 388, 539, (1997)

II.1.1.3. : Vaccinology

« Les nouveaux vaccins », Biofutur, n° 274, (Fév. 2007)
« Vaccin et vaccinologie: l'œuf et la poule », Pasteur, Le Mag, n° 2, (juin 2007)
« La cellule dendritique, un rouage essentiel de la réponse immunitaire », *in* « Vaccin et vaccinologie », Pasteur Le Mag, n° 2, p. 15, (2007)
« Vaccines : innovation and human health », EASAC policy report, (mai 2006)
Cohen J. : « SIDA : en attendant le vaccin », *in* Biofutur « Le péril infectieux, quelles stratégies de lutte ? », n° 217, (2001)

● *The challenges posed by AIDS, malaria and tuberculosis*
« Menaces sur la santé. Alerte aux virus », Journ. CNRS, n° 208, p. 22, (2007)

Camus E., Tiemoko Traoré M., Cuny G. et Aumont G. : « Le retour des maladies animales », *op. cit.*, p. 16

Doherty P.C., Zinkernagel B.M. : "A biological role for the major histocompatibility antigens", lancet, 1406-1049 (1975)

Dumbo Ogobara K. : « It takes a village : Medical research and ethics in Mali », Science, 307, 679, (2005)

Murgue B. et Robert V. : « La menace des maladies émergentes infectieuses », *in* La Recherche « Recherche pour le développement : un enjeu mondial », n° 406, p. 15, (2007)

Vogel G. : « Against all odds, victories from the front line », Science, New series, 290, n° 5491, p. 431, (2000)

Women Health Education Program
« SIDA, le combat sans répit », Journ. CNRS, n° 218, (mars 2008)

II.1.1.4. ZOONOSES

« Virus émergents. La Science sur le qui-vive » *in* Research (magazine de l'espace européen de la recherche), n° 53, p. 6, (2007)

Blancou J. et Lefèvre P.C. : « Formation à la surveillance et au contrôle des zoonoses en Afrique – Suggestions de modalités pratiques », GID, Académie des sciences, (17 mars 2006)

Brand C.M. and Skehel J.J. : « Crystalline antigen from the influenza virus enveloppe », Nature, New Biology, 238, 145, (1972)

Ducatez M.F., Olinger C.M., Owoade A.A., De landtsheer, Ammerlaan, Niesters H.G. and Osterhaus A.D. : « Avian flu, multiple introductions of H5N1 in Nigeria », Nature, 442, 37, (2006)

Gamblin S.J., Haire L.F., Russel R.J., Stevens D.J., Xiao B., Ha Y., Vasisht N., Steinhauer D.A., Daniels R.S., Elliot A., Wiley D.C. and Skehel J.-J. : « The structure and receptor binding properties of the 1918 influenza hemaglutinin », Science, 19, 1838, (2004)

Kulken T., Leighton F.A., Fouchier R.A., Le Duc J.W., Peiris J.-S., Schudel A., Stohr K. and Osterhaus A.D. : « Public health pathogen surveillance in animals », Science, 309, 1680, (2005)

Van den Hoogen B.G., De Jong J.C., Groen J., Kulken T., De Groot R., Fouchier R.A. and Osterhaus A.D. : « A newly discovered human pneumovirus isolated from young children with respiratory tract disease », Nature (Med.), 7, 719, (2001)

Wiley D.C., Wilson I.A. and Skehel J.J. : « Structural identification of the antibody binding sites of Hong Kong influenza hemagglutinine and their involvement in antigenic variation », Nature, 289, 366, (1981)

II.1.1.5. DIAGNOSIS AND THERAPY OF VIRAL DISEASES – AN OVERVIEW

• *Antiviral therapies*
• *Prion diseases*
Alper T. *et al.* : « Does the agent of scrapie replicate without nucleic acid », Nature, 214, 764, (1967)
Baumann *et al.* : « Lethal recessive myelin toxicity of prion protein lacking its central domain », EMBO J., 26, 538 (2007)
Blattler T., Brandner S., Raeber A.J., Klein M.A., Voigtlander T., Weizzmann C. and Aguzzi A. : « PrP expressing tissue required for transfer of scrapie infectivity from spleen to brain », Nature, 389, 69, (1997)
Cathala F., Brown P., Castaigne P. and Gajdusek D.C. : « La maladie de Creutzfeldt-Jacob en France continentale – Etude rétrospective de 1968 à 1977 », Rev. Neurol., 5, 439, (1979)
Cuillé J. et Chelle P.L. : « Pathologie animale : la maladie dite de la tremblante du mouton est-elle inoculable ? », C.R. Acad. Sci., 203, 1552, (1936)
Fournier J.G., *in* « Repères », *op.cit.*, (sept. 2001)
Heppner F.L. *et al.* : « Experimental autoimmun encephalomyelitis repressed by microglial paralysis », Nature (Med), 11, 146, (2005)
Heppner F.L., Musahi C., Arrighi I., Klein M.A., Rulicke T., Oesh, B., Zinkernagel R.M., Kalinke U. and Aguzzi A. : « Prevention of scrapie pathogenesis by transgenic expression of anti-prion protein antibodies », Science, 294, 5540, 178, (6 septembre 2001)
Lopez Garcia F., Zahn R., Riek R. and Wütrich K. : « NMR structure of the bovine prion protein », Proc. Natl. Acad. Sci., 97, 8334, (2000)
Meier P., Genoud N., Prinz M., Maissen M., Rulicke T. Rubriggen A., Raeler A.J. and Aguzzi A. : « Soluble dimeric prion protein binds PrP (Sc) in vivo and antagonizes prion disease », Cell, 203, 49, (2003)
Prusiner S.B. : « Prions », Proc. Nat. Acad. Sci., USA, 95, 13363, (1998)
Will R.G. *et al.* : « A new variant of Creutzfeldt-Jacob disease, in the UK », Lancet, 347, 921, (1996)
Zigas V. and Gajdusek D.C. : « Degenerative disease of the central nervous system in New Guinea endemic occurrence of kuru in the native population », N. Engl. J. Med., 257, 974, (1957)

Periodicals

Dormont D. : « Biologie des agents transmissibles non conventionnels ou prions », *in* « Revue neurol. », 154, n° 2, 142, (1998)

Le Guyader M.F.C. : « Le défi des maladies à prions », *in* Document d'information scientifique (INSERM), (septembre 2001)
Lledo P.M. : « Les maladies à Prions », PUF, collection Que sais-je ?, (2002)

II.1.2. Genetic diseases – Gene therapy

Dintzis S.M. *et al.* : « Quantitative amplification of genomic DNA from histological tissue sections after staining with nuclear dyes and laser capture microdissection », J. Mol. Diagn., 3, 22, (2001)
Fardeau M. : « L'homme de chair », Odile Jacob., (2005)
Mandel J.-L. : « Gènes et maladies. Les domaines de la génétique humaine sur la myopathie de Duchenne », Collège de France, Leçon 179, Fayard, (2005)
Quilly P. : « Duchenne de Boulogne », J.-B. Baillère et fils, Paris, (1936)

II.1.2.2. THE EXAMPLE OF DUCHENNE MUSCULAR DYSTROPHY (DMD) – A SCHOOL CASE

Ahn A.H. and Kunkel L.M. : « The structural and functional diversity of dystrophin ». Nat. Genet. 3, 283, (1993)
Monaco A.P., Bertelson C.J., Middlesworth W., Colletti C.A., Aldridge J., Fischbeck K.H., Bartlett R., Perricack-Vance M.A., Roses A.D. and Kunkel L.M. : « Detection of deletions spanning the Duchenne muscular dystrophy locus using a tightly linked DNA segment », Nature, 316, 842, (1985)

II.1.2.3. NEUROLOGICAL AFFECTIONS

« Les maladies neurodégénératives », Ann. Inst. Past., Actualités, Elsevier, 11, n° 2, (2000)
Alzheimer A. : « Centralblatt für Nervenheilkunde und Psychiatrie », 30, 177, (1907)
Brassat D., Durr A., Agid Y., Brice A. *et al.* : « Génétique de la maladie de Parkinson », Rev. Med. Interne, 20, 709, (1999)
Checler F. : « Presenilius, structural aspects and post-translational events in normal ageing and Alzheimer's disease », Molec. Neurobiol., 19, 255, (1999)
Delacourte A., David J.P., Sergeant N., Buee L., Wattez A., Vermersh P., Ghozali F., Fallot-Bianco C., Pasquier F. *et al.* : « The biochemical pathway of neurofibrillary degeneration in aging and Alzheimer disease », Neurology, 52, 1158, (1999)
Gimenez Y., Ribotta M. et Privat A. : « Sclérose latérale amyotrophique », *op. cit.*, p. 69

Hauw J.J., Dubois B., Verny M. et Duyckaerts C. : « La maladie d'Alzheimer »,
    John Libbey Eurotext, (1997)
Jellniger K.A. : « Post mortem studies in Parkinson's disease : is it possible to
    detect brain areas for specific symptoms ? », J. Neural. Transm., suppl.
    6, 1, 29, (1999)
Kingsbury A.F., Marsden C.D. and Foster O.J.F. : « DNA fragmentation in
    human substantia migra: apoptosis or perimostem effect ? », Mov.
    Disord., 13, 877, (1998)
Lebre A.S. and Brice A. : « Maladies par expansion de polyglutamine (données
    moléculaires et physiopathologiques) », in Ann. Inst. Past., Acutalités,
    11, p. 47, (2000), (inclus : Huntington, ataxies de Friedreich, retard
    mental lié à l'X, etc.)
Pericak-Vance M.A. *et al.* : « Complete genomic screen in late onset familial
    Alzheimer disease – Evidence for a new locus on chr.12 », JAMA, 278,
    1237, (1997)
Ruberg M. : « Maladie de Parkinson, vers un mécanisme de mort neuronale »,
    in Ann. Inst. Past., actualités « Les maladies neuro-dégénératives »,
    Elsevier, 11, p. 25, (2000)
Spacey S.D. and Wood N.W. : « The genetics of Parkinson's disease », Curr.
    Opin. Neurol., 12, 427, (1999)
Terry R.D. and Hansen L.A. : « Some morphometric aspects of Alzheimer
    disease and of normal aging », *in* Terry R.D. ed. « Aging and the brain »,
    Raven Press, p. 109, (1988)
Vassar R., Bennett B.D., Babu Khan S., Mendiaz E.A., Dené P., Teplow D.B.,
    Ross S., Amarante P. *et al.* : « Beta-secretase cleavage of Alzheimer's
    amyloid precursor protein by the transmembrane Aspartic-Protease,
    BACE », Science, 286, 735, (1999)
Zubenko G., Winwood F., Jacobs B., Teply I., Stiffer J., Hughes H.R., Huff F.,
    Sunderland T. and Martinez A. : « Prospective study of risk factors
    for Alzheimer's disease: results at 7.5 years », Am. J. Psychiatry, 50,
    (1999)

**II.1.2.4. SUSCEPTIBILITY GENES – POLYMORPHISMS AND DISEASES – HLA
GENES**

« Pharmacogénétique », Biofutur, 206, p. 88, (décembre 2000)
Lewis K. : « SNP's as windows on evolution », the Scientist, 16, 16, (2002)
Miki Y. *et al.* : « A strong candidate for the breast and ovarian cancer suscep-
    tibility gene, BRCA1 », Science, 266, 66, (1994)
Roses A.D. : « Pharmacogenetics and the practice of medicine », Nature, 405,
    857, (2000)
Woostee R. *et al.* : « Identification of the breast cancer susceptibility gene,
    BRCA2 », Nature, 378, 789, (1995)

II.1.2.5. GENE THERAPY – THE GENE AS A DRUG AND GENE SURGERY

Anderson F.W. : « Human gene therapy », Nature (supp.), 392, 25, (1998)

Blaese R.M., Culver K.W. *et al.* : « T. Lymphocyte directed gene therapy for ADA-SCID: initial trial results after 4 years », Science, 270, 475, (1995)

Cavazzana-Calvo M., Hacein-Bey S., de St Basile G. *et al.* : « Gene therapy of human severe combined immunodeficiency (SCID), XI disease », Science, n° 288, 669, (2000)

Fischer A. and Cavazzana-Calvo M. : « Whither gene therapy ? », The Scientist, 20, p. 36-38, (2006)

Fischer A., Cavazzana-Calvo M. *et al.* : « Sustained correction of human X linked. Severe combined immunodeficiency by ex vivo gene therapy », New Engl. J. Med., 346, 1185, (2002)

Garcia L. : « Rescue of dystrophic muscle through U7 snRNA-mediated exon skipping », Science, 306, 1796, (2004)

Hacien-Bey-Abina S., van Kalle C., Schmidt M. *et al.* : « LMD2 associated clonal T cell proliferation in two patients after gene therapy for SCID-X1 », Science, 302, 415, (2003)

Jordan B. : « Thérapie génique : espoir ou illusion ? », Odile Jacob, (2007)

II.1.3. Stem cells and cell therapy (a hope in the field of degenerative diseases)

II.1.3.1. DEVELOPMENTAL BIOLOGY CONSIDERATIONS

Ameisen J.-C. : « La sculpture du vivant », le Seuil, (1999)

Buckingham M.E. : « Actin and myosin gene families, their expression during the formation of the skeletal muscle », *in* « Essays in Biochemistry », 20, 78, (1985)

Cohen J.J. : « Apoptosis », Immunology today, 14, 126, (1993)

Ghering W. : « The homeobox : a key to understanding development ? », Cell, 40, 3, (1985)

Gros F. : « Les secrets du gène », Odile Jacob – Le Seuil, (1986)

Jacob F. : « La mouche, la souris et l'Homme », Odile Jacob, (1997)

Klarsfeld A. and Revah F. : « Biologie de la mort », Odile Jacob, (2000)

Lassar A.B., Buskin J.M., Lockshon D., Davis R.L., Apone S., Hauschka S.D. and Weintraub H. : « MyoD is a sequence specific binding protein requiring a region of myc homology to bind to the muscle creatine kinase promotes », Cell, 58, 823, (1989)

Lawrence P.A. : « The cellular basis of segmentation insects », Cell, 26, 3, (1981)

Le Douarin N.M. : « Des chimères, des clones et des gènes », Odile Jacob, (2000)
Le Douarin N.M. : « The neural crest », Cambridge Univ. Press, (1982)
Morata G. and Lawrence P.A. : « Homeotic genes, comportment and all determination Drosophila », Nature, 265, 211, (1977)
Spierer P. and Goldschmidt-Clermont : « La génétique du développement de la mouche », La Recherche, 16, 453, (1985)
Wolpert L. : « Positional information and spatial patterns of cellular differentiation », J. Theoretic Biol., 25, 1, (1969)

II.1.3.2. ADULT STEM CELLS

• *Blood stem cells*
Coulombel L. : « Les cellules souches adultes et leurs potentialités d'utilisation en recherche et en thérapeutique, cellules souches hématopoïétiques et cellules souches mésenchymateuses », Rapport au Ministre de la Recherche, p. 15, (2001)
Peault B., Oberlin E. and Tavian M. : « Emergence of hematopoietic stem cells in the human embryo. », C.R. Biologies, 325, 1021, (2002)
Prockop D.J. : « Marrow stromal cells as stem cells for non hematopoietic tissues », Science, 276, 71, (1997)

• *Other types of adult stem cells*
Montarras D. *et al.* : « Direct isolation of satellite cells for skeletal muscle repair », Science, 309, 2064, (2005)
Peschanski M. : « Les cellules souches adultes et leurs potentialités d'utilisation en recherche et en thérapeutique, comparaison avec les cellules souches embryonnaires », *in* « Neurones fétaux », Rapport au Ministre de la Recherche, p. 44, (2001)
Rochat A., Kobayashi K. and Barrandon Y. : « Location of stem cells of human hair follicles by analyses », Cell, 76, 1063, (1994)

• *Neural Stem Cells*
Gage F.H. : « Mammalian neural stem cells », Science, 287, 1433, (2000)
Mc Kay R. : « Stem cells in the central nervous system », Science, 276, 66, (1997)
Temple B. : « The development of neural stem cells », Nature, 414, 112, (2001)

• *"Plasticity" of Adult Stem Cells*
Bjornson C.R., Rietzo R.L., Reynolds B.A., Magli M.C. and Vescov A.L. : « Turning brain into blood : a hematopoietic fate adopted by adult neural stem cells in vivo », Science, 283, 534, (1999)

Blau H.M. and Blakely B.T. : « Plasticity of cell fate: insights from heter-okaryons », Semin. Cell. Dev. Biol., 10, 267, (1999)

Ferrari G., Cusella de Angelis G., Coletta M. et al. : « Muscle regenerated by bone marrow-derived progenitor », Science, 279, 1528, (1998)

Lagasse E., Connors H., Al Dhalimy et al. : « Purified hematopoietic stem cells can differentiate into hepatocytes in vivo », Nat. Med., 6, 1229, (2000)

Pettersen B.E., Bowen W.C., Patrene K.D. et al. : « Bone marrow as a potential source of hepatic oval cells », Sciences, 284, 1168, (1989)

Wilmuth I, Schieke A.E., Mc Whir J., Kind A.J. and Campbell K.H. : « Viable offspring derived from fetal and adult mammalian cells », Nature, 385, 810, (1997)

Woodbury D., Schwartz E., Prockop D. and Black I. : « Adult rat and human bone marrow stromal cells can differentiate into neurons », J. Neurosc. Res., 61, 3264, (2000)

II.1.3.3 EMBRYONIC STEM CELLS

• *Historical aspects*

Capecchi M.R. : « High efficiency transformation by direct mices infection of DNA into cultured mammalian cells » Cell, 22, 479, (1980)

Evans M.J. and Kaufman M.H. : « Establishment in culture of pluripotent cells from mouse embryos », Nature, 292, 154, (1981)

Evans M.J. : « Origin of mouse embryonal carcinoma cells and the possibility of their direct isolation into tissue culture », J. Reprod. Fert., 62, 625, (1981)

Kappler S. : « Ciblage de gène: une avancée dans la compréhension du vivant », in Découverte, n° 355, 50, (2008)

Martin G. : « Isolation of a pluripotent cell line from early mouse embryos cultured in medium conditioned by teratocarcinoma cells », Proc. Nat. Acad. Sci. (USA), 78, 7634, (1981)

Mintz B. and Illmensee K. : « Normal genetically mosaic mice produced from malignant teratocarcinoma cells », Proc. Nat. Acad. Sci. (USA), 72, 3585, (1975)

Stevens L.C. : « Experimental production of testicular teratomas in mice », Proc. Nat. Acad. Sci. (USA), 52, 654, (1964)

Stevens L.C. : « The development of transplantable teratocarcinoma from intra-testicular grafts of pre-and post fertilization mouse embryos », Dev. Biol., 21, 364, (1970)

• *Discovery of human embryonic stem cells and potential applications*

Donovan J. and Gearhardt J. : « The end of the beginning for pluripotent stem cells », Nature, 414, 92, (2001)

Le Douarin N.M. : « Thérapie cellulaire régénérative », *in* « Lettre de l'Académie des sciences », n° 20, (2006)

Shamblott M.J., Axelman J., Wang S., Bugg E.M., Littlefield J.W., Donovan P.J., Blumenthal P.D., Huggins G.R. and Gearhardt J. : « Derivation of pluripotent stem cells from cultured primordial stem cells », Proc. Nat. Acad. Sci. (USA), 95, 13726, (1998)

Thomson J.A., Itskovity Eldor J., Shapiro S.S., Waknitz M.A., Swiergiel J.J., Morshall V.S. and Jones J.M. : « Embryonic stem cell lines derived from human blastocytes », Science, 282, 1145, (1998)

• *Risks*

ISCF Ethics working pasty Letter – Oocyte donation for stem cell research, Science, 316, 368, (2007)

National Academies of Science – Assessing the medical risks of human oovyte donation for stem cell research : working report, N.A.S., Washington, USA, (2007)

Pearson H. : « Health effects of egg donation may take decades to emerge », Nature, 422, 608, (2006)

• *Somatic nuclear transfer (therapeutic cloning) – Reproductive cloning in animals*

Campbell K.H. *et al.* : « Sheep cloned by nuclear transfer from a cultured cell line », Nature, 380, 64, (1996)

Gurdon J.B., Laskey R.A. and Reeves O.R. : « The developmental capacity of nuclei transplanted from keratinized skin cells of adult frogs », J. of Embryol. and Exper. Morphol., 10, 622, (1962)

Rideout W., Egan K. and Jeanisch R. : « Nuclear cloning and epigenetic reprogramming of the genome », Nature, 293, 1093, (2001)

Wakayama T. *et al.* : « Differentiation of embryonic stem cell lines generated form adult somatic cells by nulear transfer », Science, 292, 740, (2001)

II.1.3.4. ETHICAL ASPECTS OF THE USE OF EMBRYONIC STEM CELLS

« Interspecies embryos », The Academy of Medical sciences, a report by the Academy of medical science, (juin 2007)

De Coppi P. *et al.* : « Isolation of amniotic stem cell lines with potential for therapy », Nature Biotechnology, 25, 100, (2007)

Fagniez P.-L. : « Cellules souches et choix éthiques », Rapport au Premier Ministre, la Documentation française, (2006)

Guan K. *et al.* : « Pluripotency of spermatogonial stem cells from adult mouse testis », Nature, 440, 119, (2006)

Maherall N. *et al.* : « Directly reprogrammed fibroblasts show global epigenetic remodelling and widespread tissue contribution », in press

Sureau C. : « Cellules souches embryonnaires, aspects spécifiques, médicaux et éthiques », Journée Commune Académie nationale de médecine-Académie des sciences, Bull. Acad. nat. Méd., 184, 1139, (2000)
Takahashi K. and Yamanaka S. : « Induction of pluripotent stem cells from mouse embryonic and adult fibroblast cultures by defined factors », Cell, 126, 663, (2006)
Wernig M., Meissner A., Foreman R., Brambrink T., Ku M., Hochedlinger K., Bernstein B.E. and Jaenisch R. : « In vitro reprogramming of fibroblasts into a pluripotent ES-cell-like state », Nature, 448 (7151), 318, (2007)

II.1.4. Ageing – Senescence and cell death (Apoptosis) – Cancers

II.1.4.1. AGEING – GENERAL CONSIDERATIONS

Gros F. : « Sciences et pays en développement : l'Afrique subsaharienne francophone », Rapport RST n° 21 de l'Académie des sciences, EDP Sciences, (2005)
Léridon H. : « Colloque de l'Académie des sciences sur le vieillissement », (2005)

II.1.4.2. GENETICS AND LONGEVITY

Baulieu E.E. : « Génération pilule », Odile Jacob, (1990)
Bernard J. : « Vieillir, Entretiens avec Antoine Hess, » Calmann-Lévy, (2001)
Friedman D.B. and Johnson T.E. : « A mutation in the age-1 gene in *Caenorhabditis elegans* lengthens life and reduces hermaphrodite fertility », Genetics, 118, 75, (1988)
Kimura K.D. *et al.* : « *daf-2*, an Insulin Receptor-Like Gene That Regulates Longevity and Diapause in *Caenorhabditis elegans* », Science, 277, 942, (1997)
Larsen P.A., Albert P.S. and Riddle D.L. : « Genes that regulate both development and longevity in *Caenorhabditis elegans* », Genetics, 139, 1567, (1995)
Morri J.Z., Tissenbaum H.A. and Ruvkun G. : « A phosphatidylinositol-3-OH kinase family member regulating to diapause in *Caenorhabditis elegans* », Nature, 382, 536, (1996)

• *Relationships between genomics and longevity in the human species*
Edelstein S.J. : « Gènes et longévité », *in* « des gènes aux génomes », p. 85-90, Odile Jacob, (2002)
Gros F. : « Génétique, sénescence et mort », *in* Mémoires scientifiques, un demi-siècle de Biologie, Odile Jacob, p. 270-282, (2003)
Pecca A.A. *et al.* : « A genome wide scan for linkage to human exceptional longevity identifes a locus on chromosome 4 », Proc. Nat. Acad. Sci. (USA), 98, 10505, (2001)

- *The causes of physiological ageing*
- *Molecular ageing – effects of free radicals*
Landousy M.T. : « Le protéasome et ses inhibiteurs », Biofutur, n° 243, 11, (2004)

II.1.4.3. CELLULAR SENESCENCE

Kim N.W. *et al.* : « Specific association of human telomerase activity with immortal cells and cancer », Science, 266, 2011, (1994)

II.1.4.4. APOPTOSIS – PROGRAMMED CELL DEATH

Ameisen J.C. : « La sculpture du vivant: le suicide cellulaire ou la mort créatrice », le Seuil, (1999)
Ameisen J.-C. : « Apoptose en pathologie humaine », *in* Ann. Inst. Past. Actualités, 11, n° 4, (2000)
Auboine G., *in* « Biologie et Géologie », APBG, n° 3, 485, (2003)
Bialik S. and Kimchi A. : « The death associated protein kinases: structure, function and beyond », Ann. Rev. Biochem., 75, 189, (2006)
Duboule D. et Sordino P. : « Des nageoires aux membres : l'apport de la génétique moléculaire du développement dans l'étude de l'évolution de la morphogénèse chez les vertébrés », Médecine/Sciences, 12, 147, (1996)
Ellis R.E., Yuan J.Y. and Horvitz H.R. : « Mechanisms and functions of cell death », Ann. Rev. Cell. Biol., 7, 663, (1991)
Kerr J.-J. *et al.* : « Apoptosis a basic biological phenomenon with wide ranging implications in tissue kinetics », Brit. J. Cancer, 26, 239, (1972)
Llambi F., Causeret F., Bloch-Gallego E. and Mehlen P. Llambi F., Causeret F., Bloch-Gallego E. and Mehlen P. : « Netrin-1 acts as a survival factor via its receptors UNC5H and DCC », EMBO J., 20, 2715, (2001)

Periodical

Jacotot E., Ferri K.F. and Kroemer G. : « Apoptose et mitochondries: le côté obscur de l'organite », Ann. Inst. Past. Actualités, 11, 19, (2000)

II.1.4.5. APOPTOSIS AND CANCER

Gross A. : « BTD as a double agent in cell life and death », Cell cycle, 5, 582, (2006)
Kamer I. *et al.* : « Proapoptotic BID is an ATM effector in the DNA damage response », Cell, 122, 593, (2005)
Mazelin L. *et al.* : « Netrin-1 controls colorectal tumorigenesis by regulating apoptosis », Nature, 431, 80, (2004)
Reed J.C. : « Dysregulation of apoptosis in cancer », J. Clin. Oncol., 17, 2941, (1999)

Système Fas/Fas L (voir Ann. Inst. Past. Actualités, Apoptose en pathologie humaine)

Thiebault K., Mazelin L., Pays L., Joly M.O., Scoazec J.Y., Saurin J.C., Romeo G. and Mehlen P. : « The netrin-1 receptors UNC5H are putative tumor suppressors controlling cell death commitment », Proc. Natl. Acad. Sci. (USA), 7, 4173, (2003)

II.1.4.6. MOLECULAR MECHANISMS OF APOPTOSIS

Ferry L. et Vincent J.-D. : « Qu'est-ce que l'homme ? », Odile Jacob, (2001)

Gross A. : « Mitochondrial carrier homolog 2 : a clue to cracking the BCL-2 family riddle ? », Bioenergetics and Biomembranes, 37, 113, (2005)

Jacobson M.D., Weil M. and Raff M.C. : « Programmed cell death in animal development », Cell, 88, 347, (1997)

Verhagen A., Ekert P.G., Pakusch M., Silke J., Connolly L.M., Reid G.E., Moritz R.L., Simpson R.J. and Vaux D.L. : « Identification of DIABLO a mammalian protein that promotes apoptosis by binding to and antagonizing IAP proteins », Cell, 102, 43, (2000)

II.1.4.7. CANCERS

• *Epidemiological facts*

Blaudin de Thé G. : *in* « Sciences et Pays en développement : l'Afrique subsaharienne francophone », RST n° 21, EDP Sciences, p. 103, (2006)

Blaudin de Thé G. : « Sur la piste du cancer », Flammarion, (1984)

Burkitt D.P. : « A sarcoma involving the jaws in African children », Brit. J. Surg., 46, 218 ; (1958)

Ferloy J., Black R., Whelan S.L. and Parkin D.M. : « Cancer incidence in five continents », VIII, IARC, Scientific publication, n° 155, (2003)

Hardford J.B. : *in* « Changing lives », Biovision-Alexandria, (2006), éds. I.S. Serrageldin, E. Masood, M. El Fahani et A. Massoud, p. 137, (2007)

Jemal A., Thomas A., Mauray T. and Thun M. : « Cancer statistics », CA, Cancer. J. Clin., 23, (2002)

Khayat D. : « Les chemins de l'espoir », Odile Jacob, (2003)

Organisation mondiale de la santé : « World Health Statistics Annual », Genève, OMS, (1996-1998)

• *Biology of cancer – oncogenes – suppressor genes – repair system*

« Oncogenes and the molecular origine of cancers », Cold. Spring. Harb., Lab. Press, 327, (1989)

Amundson S.A., Myers T.G. and Fornace A.J. : « Roles for p53 in growth arrest and apoptosis: Putting on the brakes after genotoxic stress », Oncogenes, 17, 3287, (1998)

Bishop J.M. : « Retrovirus and Oncogenes II », *in* les prix Nobel, Almqvist and Wiksell, 220, (1989)

Bishop J.M. : « Oncogenes and clinical cancer », in Weinberg, édit. R.A.

Hanahan D. and Weinberg R.A., « The hall marks of cancer », Cell, 100, 57, (2000)

Kastan M.B., Onyekewere O., Sidransky D., Vogelstein B. and Craig R.W. : « Participation of p53 protein in the cellular response to DNA damage », Cancer Res, 51, 6304, (1991)

Levine A. : « The cellular gate keeper for growth and division », Cell, 88, 53, (1997)

Monier R. : « Aspects fondamentaux: mécanismes de cancérogénèse et relation dose-effet », C.R. Acad. Sci., 323, 603, (2000)

Stehelin D., Varmus H.E., Bishop J.M. and Vogt P.K. : « DNA related to transforming gene(s) of avian sarcoma viruses is present in normal avian DNA », Nature, 260, 170, (1976)

Varmus H.E. : « An historical overview of oncogenes », *in* Weinberg R.A. édit., « Oncogenes and the molecular origins of cancer », Cold. Spring. Harb., Lab. Press, 3, (1989)

Varmus H.E. : « Retroviruses and oncogenes I », *in* Les prix Nobel, Almqvist and Wiksell, 194, (1989)

• *Repair systems and cancers*

« Le réparosome », Biofutur, 271, 17-48, (2006)

Boiteux S., Castaing B. et Radicella P. : « BER : mécanismes moléculaires et rôles biologiques », Biofutur, 271, 35, (2006)

• *Epigenetic factors*

Feinberg A.P. and Vogelstein B. : « Hypomethylation distinguishes genes of some human cancers from their normal counterparts », Nature, 301, 89, (1983)

Gaudet F. *et al.* : « Induction of tumors in mice by genomic hypomethylation », Science, 300, 489, (2003)

Herman J.G. and Baylin S.B. : « Gene silencing in cancer in association with promoter hypermethylation », New. Engl. J. Med., 349, 2042, (2003)

Holliday R. and Pugh J.E. : « DNA modification mechanisms and gene activity during development », Science, 187, 226, (1975)

Jones P.A. and Baylin S.B. : « The fundamental role of epigenetic events in cancer », Nat. Rev. Genet., 3, 415, (2002)

Riggs A.D. : « X inactivation, differentiation, and DNA methylation », Cytogenet. Cell Genet., 14, 9, (1975)

• *Epigenetic control of differentiation in cancer stem cells*

Cui H. *et al.* : « Loss of IGF2 imprinting : a potential marker of colorectal cancer risk », Science, 299, 1753, (2003)

Di Croce L, Raker V.A., Corsaro M., Fazi F., Fanelli M., Faretta M. *et al.* : « Methyltransferase recruitment and DNA hypermethylation of target promoters by an oncogenic transcription factor », Science, 295, 1079, (2002)

Feinberg A.P., Ohlsson R. and Henikoff S. : « The epigenetic progenitor origin of human cancer », Nature, Rev. Genet., 7, 21, (2006)

Hochedlinger K., Blelloch R., Brennan C., Yamada Y., Kim M., Chin L. and Jaenisch R. : « Reprogramming of a melanoma genome by nuclear transplantation », Genes Dev., 18, 1875, (2004)

Kleinsmith L.J. and Pierce J.G.B. : « Multipotentiality of single embryonal carcinoma cells », Cancer Res., 24, 1544, (1964)

Levan A. and Hauschka T.S. : « Endomitotic reduplication mechanisms in ascites tumors of the mouse », J. Natl. Cancer Inst., 14, 1, (1953)

Li L., Conelly M.C., Wetmore C., Curran T. and Morgan J.I. : « Mouse embryos cloned from brain tumors », Cancer Res., 63, 2733, (2003)

Lotem J. and Sachs L. : « Epigenetics and the plasticity of differentiation in normal and cancer stem cells », Oncogene, 25, 7663, (2006)

Marks P.A., Rifkind R.A., Richon V.M., Breslow R., Miller T. and Kelly W.D. : « Histone deacetylases and cancer: causes and therapies », Nat. Rev. Cancer, 1, 194, (2001)

Reik W., Dean W. and Walter J. : « Epigenetic reprogramming in mammalian development », Science, 293, 1089, (2001)

Sachs L. : « The control of hematopoiesis and leukemia: from basic biology to the clinic », Proc. Natl. Acad. Sci. USA, 93, 4742, (1996)

Surani M.A. : « Reprogramming of genome function through epigenetic inheritance », Nature, 414, 122, (2001)

II.2. AGRICULTURE – NUTRITION – FEEDING MANKIND – THE CHALLENGES OF MALNUTRITION – TRANSGENIC PLANTS (DATA, HOPES AND FEARS)

II.2.1. Feeding mankind – Data on the problem and the challenges to be met

Persly G.J., Peacock J. and Van Montagu M. : « Biotechnology and sustainable agriculture », *in* Series on Sciences for sustainable development, ICSU, n° 8, (2002)

Pinstrup-Andersen P., Plandya-Lorch R. and Rosegrand M.W., « World Food Prospects: critical issues for the Early twenty first century », International Food Policy, Research Institute, (1999)

Periodicals

« Sciences et pays en développement – Afrique subsaharienne francophone », sous la direction de F. Gros, Académie des sciences, Rapport sur les Sciences et les Technologies n° 21, EDP Sciences, (2006)

« L'eau pour tous – l'eau pour la vie – le nouveau courrier », UNESCO n° 3, (oct. 2003)

II.2.1.2. A WORLD FOOD CRISIS – THE "RETURN OF HUNGER"

Jacquet P. : « Le retour de la faim », Sciences au Sud, Le journal de l'IRD, n° 44, p. 1, (avril-mai-juin 2008)

II.2.2. Contributions of genomics

Cushman J.-C. and Bohnert H.J. : « Genomic approaches to plant stress tolerance », Current Opinion in Plant Biology, 3, 117, (2000)

Douce R. (animateur) : « Le monde végétal – Du génome à la plante entière », Académie des sciences, Rapport TST, n° 10, Tec et Doc, (2000)

Durand-Tardif M., Candresse Th. : « Apport de la génétique à la protection des plantes », Biofutur, 242, 16, (mars 2004)

Glazenbrook J. : « Genes controlling expression of defence responses in Arabidopsis », Current opinion in Plant Biology, 2, 280, (1999)

Kush G.S. : « Green revolution preparing for the 21st century », Genome, 42 (4), 646, (1999)

Sasson A. and Elliott M.C. : « Agricultural biotechnology for developping countries: a strategic overview », *in* Christou P. and Klee A. édit., H. Handbook of Plant biotechnology, John Wiley and sons, p. 2001, (2004)

II.2.3. Transgenic plants – Some general data

Badr E. : « GM crops, Food security and the environment », *in* « Changing lives », Biovision Alexandria, (2006), I. Serrageldin and E. Méasood eds with M. El-Fahan and A. Massoud, Bibliotheca Alexandrina Cataloging in publication data, p. 217, (2007)

Clive J. : « A decade of Agricultural Biotechnology », *in* Changing lives, *op. cit.*, p. 235, (2007)

Fernandez-Cornjio J. and Cawell M. : « The first decade of genetically engineered rops in the United States », USDA Economic information bulletin, n° 11, p. 1-36, available at : http//www.ers.usda.gov

Persley G. and Siewdow J. : « Applications of Biotechnology to crops: benefits and risks », *in* Ruse M. and Castle E. edit., Genetically modified foods, Prometheus Books, p. 221, (2002)

• *Principal types of modification introduced by plant transgenesis with agricultural aims*
Clive J. : « Tolérance aux herbicides », *op. cit.*, p. 237

• *Drought et salinity*
AGERI 2005 : « Broding triticum durum in Mediterranean region by using in vitro and genetic transformation tools », available at http//www.ageri. sci.eg/topic6/durum.htm
CIMMIYT 2003, « Drought relief, Seed Relief in Sight », available at http// www.cimmyt.cgiar.org/whatiscimmyt/recentar/Dsupport/drought.htm
CIMMY 2004, « Molecular approaches for the genetic improvement of cereals for stable production in water limited environment », J.-M. Ribaud and D. Poland eds, (2004)
FAO 2003 : « Report of the FAO expert consultation on Environmental effects of genetically modified crops », FAO, (16-18 June 2003)
Garg A.K., Kim J.K., Owens T.G., Ranvala A.P., Choi Y.D., Kocian L.V. and Wu R.J. : « Trehalose accumulation in rice plants confers high tolerance levels to different abiotic stress », Proc. Nat. Acad. Sci., USA, 99, 15 898, (2002)
Johnson B. : « Potential environmental impacts form Novel Crops », *in* Changing lives, op. cit., p. 224, (2007)
Madkour M. : « Harnessing new science to meet the challenges of Drought », *in* Changing lives, Op. Cit., p. 299, (2007)
Su J., Shen Q., Ho T. H.D. and Wu R. : « Dehydration-stress-regulated trans-gene expression in stably transformed rice plants », Plant Physiol., 117, 913, (1998)

• *Other characteristics*
Hayakawa, Zhu Y., Ito K, and Kimura Y. : « Genetically engineered rice, resistant to rice stripe virus, an insect transmitted virus », Proc. Natl. Acad. Sci. USA, 89, 9865, (1992)
Nena Vishvanath : « Protecting live stock through genomics », in changing lives, *op. cit.*, p. 241, (2007)

• *Overall physiology – Nutritional value*
Abbott J.C., Barakate A., Pincon G., Legrand M., Lapierre C., Mila I., Schuch W. and Halpin C. : « Simultaneous suppression of multiple genes by single transgenes – Down regulation of three unrelated lignin biosynthesis genes in tobacco », Plant Physiol., 128, 844, (2002)
Cobbett C.S. : « Phytochelatins and their roles in heavy metal detoxification », Plant Physiology, 123, 825, (2000)

Ku M.S.B. *et al.* : « High levels expression of maize phosphoenol pyruoate carboxylase in transgenic rice plants », Nature, Biotechnology, 17, 76, (1999)

Mann C.C. : « Genetic engineers aim to soup up crop photosynthesis », Science, 283, 314, (1999)

Matsuoka *et al.* (2001), cité dans l'article de Elliott M., Sasson A. et Cockburn A. : « Starvation, obesity or optimized diets : which way for nutrition ? », *in* Changing lives, *op. cit.*, p. 249, (2007)

Stark D.M. *et al.* : « Regulation of the amount of starch in plant tissues by ADP glucose pyrophosphorylase », Science, 285, 287, (1992)

• *Plant transgenesis and health*

Acharya T., Daar A.S., Thorsteindottir H., Dowdeswell E. and Singer P.A. : « Strenghtening the role of genomics in global Health », Plos Medicine, 1, n° 3, E-40, (2004)

Falco S.C. *et al.* « Transgenic canola and soybean seeds with increased lysine », Biotechnology, 13, 577, (1995)

Hoa *et al.* (2003), cité dans « Elliott M., Sasson A. et Cockburn A. », *in* Changing lives, (2007)

Larrick J.W. and Thomas D.W. : « Producing proteins in transgenic plants and animals », Current Opinion in Biotechnology, 12, 411, (2001)

Ye X.S. *et al.* : « Engineering the provitamin A (beta carotene), biosynthetic pathway into (carotenoid-free) rice endosperm », Science, 287, 303, (2000)

• *Hopes Reservations – Potential mankind*

FAO, 2003b : « Report of the FAO Expert, consultation on Environmental effects of Genetically modified crops », FAO, (June 2003)

Johnson B.R. : « Gene flow from crops to crops and form crops to wild relatives – does it matter ecologically ? », Aspects of Applied Biology, 74, G. Crops, Ecological dimension, eds. Van Emlen, H.F. *et al.*, p. 53, (2004)

Khush G.S. and Ma J. : « Crop biotechnology for developing countries : opportunity and duty », *in* Christon P. and Klee H. (eds), Handbook of plant Biotechnology, John Wiley and Sons, p. 1313, (2004)

Rousoh R. : « Can we stop adaptation pests to insect transgenic crops ? », *in* Biotechnology and integrated pest management, G. Persley, CABI internat., p. 242, (1996)

Sasson A. and Elliott M.C. : (cf. référence *in* II.2.3.)

Swaminathan M.S. : « An evergreen revolution », Biologist, 47, n° 2, 85, (2000)

Van Montagu M. and Bursens S. : « From molecular genetics to Plants for the future », Library of Alexandria conference on Biotechnology and sustainable development : voices of the South and North, (March 2002)

II.3. ENVIRONMENT – ENERGY – BIODIVERSITY

II.3.1. Energy challenges – Greenhouse effects – Renouwable energies – Biofuels

II.3.1.1. ENERGY CHALLENGES – CLIMATE CHANGE

« Évolution des climats », la Lettre de l'Académie des sciences, n° 21, (2007)

« Remous sur les biocarburants », *in* RDT-Info, Magazine de Recherche européenne, p. 30, (2006)

Ballerini D. : « les Biocarburants », *in* « Anticiper les ruptures », Découverte *op.cit.*, p. 66, (2007)

Estèves B. and Almeida C. : « Les Biocarburants en question », Biofutur, 288, 48, (2008)

Fagione J. *et al.* : « Land clearing and the biofuel carbon debt », Science, 319, 1235, (2008)

« Réchauffement climatique : impact sur les maladies infectieuses à vecteur », *in* ECRIN, Recherche-Technologie, Société, n° 68, p. 22, (2007)

Sales C. : « Énergie ; les promesses de la biomasse », *in* La Recherche, n° 406, Recherche pour le développement un enjeu mondial, p. 24, (2007)

Searchinger T. *et al.* : « Use of US croplands for biofuels increases greenhouse gases through emissions from land-use change », Science, 319, 1238, (2008)

Solomon I. et Gros F. : « Énergie solaire et santé dans les pays en développement », Acad. Sci., Colloque COPED-UNESCO, Tec et Doc Lavoisier, (2004)

Tissot B., *in* « Énergie 2007-2050, Les choix et les pièges », Académie des sciences, B. Tissot, 10, (2007)

Varène V. : « Énergies – anticiper les ruptures », *in* Découverte, Revue du Palais de la Découverte, n° 344-345, p. 10, (Janvier-février 2007)

II.3.2. Biodiversity

II.3.2.1. IDENTIFYING AND PROTECTING BIODIVERSITY

II.3.2.1.1. General data – Threats and concerns for a common heritage

Bastos Da Veiga J. *et al.* : « La longue marche de l'Agriculture durable en Amazonie », *ibid.*, p. 28, (2007)

Boisvert V., Bazile D. : « Biodiversité, des codes de bonne conduite », *in* La Recherche, n° 406, p. 26, (2007)

Coquart J. : « Menaces sur la Biodiversité – Les réponses de la Science », Journal du NCRS, n° 180, 18, (2005)

Gilman N., Randall D. and Schwartz P. : « Impacts of climat change », Global Business Network, (juin 2007)

Lévèque C. et Mounolou J.-C. : « Biodiversité – Dynamique biologique et conservation », Masson Sciences, Dunod, (2001)

Marjorie L., Reaka-Kudla W., Don E. and Wison E.O. : « Understanding and protecting our biological ressources », Biodiversité II, Joseph Henry Press, (1997)

Testard Vaillant P. : « 5 défis pour la biodiversité », Journal du CNRS, n° 196, 18-27, (2006)

Wilson E.O. : « La diversité de la vie », Odile Jacob, (1993)

II.3.2.1.2. The variety of living species – An unfinished investigation

Académie des sciences : « Biodiversité et environnement », Rapport n° 33, (1995)

Académie des Sciences : « Systématique ; ordonner la biodiversité », Rapport sur la Science et la Technologie, n° 11, Tec et Doc, (2000)

II.3.2.1.3. Phylogenetic relations – Genomic comparisons

De Long E.F. : « Archaea in coastal marine environments », Proc. Natl. Acad. Sci. (USA), 89, 5685, (1992)

Larsen N., Olsen G.J., Maidak B.L., McCaughey M.J., Overbeek R., Macke T.J., Marsh T.L. and Woese C.R. : « The ribosomal data base project », Nucleic Acid Res., 21 (suppl.), 3021, (1993)

Leipe D.D.O., Wainright J.H. *et al.* : « The stramenopiles from a molecular perspective: 16S – like rRNA sequences from Labyrinthuloides minuta and Cafeteria roenbergensis », Phycologia, 33, 369, (1994)

Sogin M.L. and Hinkle G. : « Common measures for studies of Biodiversity: molecular phylogeny in the eukaryotic microbial world », in Biodiversity II, Understanding and protecting our biological ressources, *op. cit.*, chapter 8, p. 109, (1997)

Sogin M.L. : « Evolution of eukaryotic micro-organisms and their small summit ribosomal RNAs », Amer. Zool., 29, 487, (1989)

II.3.2.1.4. Plant Genomics and Biodiversity

Delseny M. : « Re-evaluating the relevance of ancestral shared synteny as a tool for crop improvement », Curent Opin. Plant Biol., 7, 126, (2004)
Quétier F., Salanoubat M. and Weissenbach J. : « Le séquençage des génomes nucléaires des plantes », *in* Biofutur, 265, 27, (2006)

II.3.2.1.5. Animal Genomics and Biodiversity

II.3.2.1.6. Biodiversity of microorganisms – Metagenomics

« Prevalence of Plant Pathogenic viruses », PLos Biology, 4, 0108, (2006)
Brettbart M. *et al.* : « Metagenomic Analyses of an Uncultured Viral Community from Human Feces », J. of Bactériol., 185, 6220, (2003)
Delong E.F. : « Microbial community genomics in the ocean », Nature Rev./Microbiol., 3, 459, (2005)
Fraser C.M. *et al.* : « Complete genome sequence of Treponema pallidum, the syphillis spinochete », Science, 281, 375, (1998)
Gewin V. : « Discovery in the dirt », Nature publishing Group, p. 384, (2006)
Hallam S.J. *et al.* : « Reverse Methanogenesis: testing the hypotesis with environmental genomics », Science, 305, 1457, (2004)
INRA-Dossier « Les Recherches de l'INRA sur la flore digestive », n° 14, (mai 2006)
Rondon M.R. *et al.* : « Cloning the soil metagenome : a strategy of assessing the genetic and functional diversity of uncultured micro-organisms », Applied and Environmental Microbiol., 66 (n° 6), 2541, (2000)
Strous M. *et al.* : « Deciphering the evolution and metabol. of an anarmox bacterium from a community genome », (Letter), Nature Publishing Group, 440, 790, (2006)
Tringe S.G. and Rubin E. : « Metagenomics: DNA sequencing of environmental samples », Nature Reviews, Genetics, 96, 805, (2005)
Venter J.C. *et al.* : « Environmental genome shotgun sequencing of the Sargasso Sea », Science, 304, 66, (2004)
Vignais P. : « Science expérimentale et connaissance du vivant », EDP sciences, p. 268, (2006)

www.ingramcontent.com/pod-product-compliance
Lightning Source LLC
Chambersburg PA
CBHW021035210326
41598CB00016B/1034